SPACE, TIME AND INCARNATION

Thomas F. Torrance

Professor of Christian Dogmatics in the University of Edinburgh

OXFORD UNIVERSITY PRESS

OXFORD LONDON NEW YORK

OXFORD UNIVERSITY PRESS
Oxford London Glasgow
New York Toronto Melbourne Wellington
Ibadan Nairobi Dar es Salaam Cape Town
Kuala Lumpur Singapore Jakarta Hong Kong Tokyo
Delhi Bombay Calcutta Madras Karachi

First published by Oxford University Press, London, 1969
First issued as an Oxford University Press paperback, 1978

The Library of Congress cataloged the first printing of this title as follows:

Torrance, Thomas Forsyth, 1913 —
Space, time and incarnation [by] Thomas F. Torrance.
London, New York [etc.] Oxford U. P., 1969.

xi, 92 p. 23 cm. 25/— B 69–14610
Bibliographical footnotes.
1. Incarnation. 2. Space and time. I. Title.
BT220.T6 232′.1 72–410734
ISBN 0–19–520082–9 pbk.

Printed in the United States of America

SPACE, TIME AND INCARNATION

PREFACE

The purpose of this little book is threefold:

(i) to lay bare the ground on which modern Protestant theology has attempted to detach the message of the Christian Gospel from any essential relation to the structures of spacc and time;

(ii) to examine the place of spatial and temporal ingredients in basic theological concepts and statements and to clarify the epistemological questions they involve;

(iii) to offer a positive account of the relation of the incarnation to space and time, by penetrating into the inner rational structure of theological knowledge and letting it come to articulation within the context of modern scientific thought.

The intention throughout is to further the clarifying, refining and deepening of these basic theological concepts without lapsing into unscientific reductionism, but with the hope of contributing to coherent and unitary understanding of the Christian Faith.

The instrument I have selected for this purpose is the question of the part played by the receptacle notion of space in the history of Christian thought. This question was forced on me when examining the problem of the so-called 'Calvinist extra' which has long been the subject of controversy between Lutheran and Reformed theologians. The putting of any fresh question to what is already very familiar can be most revealing, but the putting of this question to Ancient, Medieval, Reformation, and Modern Theology has proved particularly fruitful for the light thrown upon the problems and confusions in which we are still immersed. Only some of the results of this investigation are presented here, and for the most part shorn of the apparatus of detailed evidence, which allows the main argument to stand out in sharp relief. Those who are familiar with the history of Christian thought will not need elaborate footnotes from me, while those who are not familiar with it can check what I have to say by a study of the sources for themselves, to which I have given ample pointers. The evidence for my account of Patristic thought, however, is given in some fullness in a contribution to a *Festschrift* in honour of Professor Georges Florovsky to be published in New York in 1969.

Since a selection had to be made, the reader should be warned that large tracts in the history of ideas are not treated at all. Some will miss in particular an examination of the notion of time offered by St. Augustine and of its place in the history of Augustinian thought for a thousand years or more—the immense importance of which I would not belittle, even when the character of Western thought changed so considerably with the ascendancy of the *philosophia perennis*. What I have done is rather to examine the problem of spatial concepts in Nicene Theology, with a view to determining how the Christian Church in its decisive and formative period presented the Gospel in the context of ancient culture and science, and developed its own specifically theological understanding of the relation of God to the world of space and time, on the ground of the doctrines of creation and incarnation. That is presented here in the first chapter, which was delivered as a lecture to the Faculty of Divinity at King's College, London, in February 1968.

While the Fathers rejected the receptacle idea of space, the

Medievals introduced it into the West with very far-reaching effects. This called for an examination of the role given to it in Reformation thought, Newtonian science, and subsequent Protestant philosophy and theology, in which a deistic dualism came to be built into the fabric of Western thought. I have not attempted here to show the developments in Reformed, Anglican, or Roman theology in relation to this, but have concentrated on German Lutheran thought which has certainly made the pace for much modern theology. It is to this that the second chapter is devoted, which represents a lecture delivered to the Faculty of Theology in Manchester University in March 1968.

The third chapter comprises my presidential lecture to the *Society for the Study of Theology* delivered at its recent session held in St. Edmund Hall, Oxford, 26–8 March 1968. In this I have tried to think through the relation of the incarnation to space and time in a constructive way in the context of modern science, and at the same time to destroy the damaging dichotomies that have fostered corruption and confusion in modern theology. Admittedly this is in sharp contrast to the so-called 'new theology' of recent years, which, for all its claims, is evidently a movement of reaction against the hard, exact and progressive thinking of pure science under the control of trans-subjective realities, and is more and more the expression of the culture of a technological society in which thinking is not controlled from beyond our own personal and social constructions. Thus instead of being so far in advance of society that it really is in a position to help it, a theology of this kind becomes inevitably geared into the popular mind with its sedimentary deposit of obsolete ideas, and into the prevailing ethic which, in the present state of socio-psychopathology, is characterized so widely by regression into more infantile forms of behaviour. It is not surprising that at critical junctures in history where the Gospel and social culture are forced apart that theological statements redacted to cultural statements should appear so empty and meaningless. In face of problems such as these, this book is a plea for a scientific theology which can really help modern men refer their thoughts properly beyond themselves to the living God.

Theological science and natural science are both at work in the same world seeking understanding within the rational connections and regularities of space and time where they pursue their respective inquiries and let their thinking serve the reality into which they seek to inquire. This does not mean that theology can allow its own subject-matter to be determined by the results of scientific work in other fields or that it can extrapolate their particular procedures into its own field of operation, but that it must pursue its own distinctive ends in a scientifically rigorous way on its own ground and in accordance with the nature of its own proper object. Yet because it operates in the same world as natural science it cannot pursue its activity in a sealed-off enclave of its own, but it must take up the relevant problems and questions posed by the other sciences in clarifying knowledge of its own subject-matter. Hence it can make legitimate use of analogies taken from the other sciences where similar problems arise in order to help it penetrate into the inherent intelligibility of its own object, and under its control bring it to such precise articulation in its understanding that there is no confusion between knowing and what is known, and no unwarranted intrusion of subjective factors into the transcendental content of its knowledge.

Scientific work of this kind has considerable ecumenical significance. By distinguishing objective realities from subjective states it cuts away alien ideas and the pseudo-theological structures arising out of them which are always divisive, and so contributes to the inner coherence and unification of theology on the objective ground common to us all. Moreover, since scientific explanation involves an understanding which we can communicate to others in such a way as to command their assent, we are committed by scientific procedures to the achievement of common understanding and to control through mutually accepted explanation. The profound questioning that this involves will undoubtedly produce tension and upheaval, but, the more scientific theological activity, is the more it will make it clear that tension and upheaval can arise only when we cling irrationally to non-theological factors found in the psychological and social conditioning of our concepts. Ecumenical dialogue that brings these non-theological factors to the surface can in its turn serve rigorous scientific work in theology by revealing the impurities and false connections that need to be refined away. It is in this light that I would ask my readers to regard the critique I have been forced to offer of certain developments in German Lutheran thought—I am no less critical of my own Reformed tradition in its development in certain other directions. I am wholly convinced that the more scientifically we can pursue our theology the more we shall be able to march forward as one, and the more relevant our message will be to a world that will always be dominated by empirical and theoretical science.

I am most grateful to my son Iain for his revision of the proofs and preparation of the Index, and also for many valuable comments on the text at different stages. Once again I am happily indebted to Mr. G. N. S. Hunt and his colleagues for steering this little book so skilfully through the processes of publication.

Edinburgh, Easter, 1968. THOMAS F. TORRANCE

To

The Society for the Study of Theology

in gratitude and encouragement

CONTENTS

CHAPTER 1

THE PROBLEM OF SPATIAL CONCEPTS IN NICENE THEOLOGY

When the Nicene–Constantinopolitan Creed speaks of Jesus Christ who was incarnate by the Holy Ghost of the Virgin Mary and crucified under Pontius Pilate as 'being of one substance with the Father, by whom all things were made, who for us men and for our salvation came down from heaven', it is clearly affirming that God Himself in His own being is actively present with us as personal Agent within the space and time of our world. These credal assertions are meant to be positive statements intelligibly correlated with the self-revelation of God in Jesus Christ, enshrining concepts that have objective truth corresponding to them because they are grounded finally in the nature and activity of God Himself. Thus in spite of the hymnic character of the Creed, its language cannot be treated as if it were merely symbolic (or indeed essentially symbolic), employing aesthetic, non-conceptual forms of thought that are related to God in a detached, oblique way, but that derive their meaning and justification mainly through coordination with the religious imagination and self-understanding of the Church. Rather is the language to be regarded as essentially significative, employing conceptual forms that are intended to refer us to God in a direct and cognitive way and that have their meaning and justification precisely in that objective and operational reference. There should be little doubt about the fact that the Nicene Fathers were convinced that the disciplined statements they made in formulating the Creed were rightly and properly related to what they signified, i.e. through a basic conceptuality that did not vary with the many forms of man's own devising but was controlled by the reality intended.

This makes it all the more important for us to examine the way in which spatial language is used in such statements. If they are merely symbolic, then the spatial element in them can be interpreted quite easily, in a merely metaphorical or tropical sense, yet at the expense of any conceptual correlation with the inherent intelligibility of God. But if they are essentially significative, then the conceptual content of the statements must have some real correlation with God's own inherent intelligibility through which they fulfil their intention not only of indicating His reality but of affording us, in some measure at least, definite cognitive apprehension of God in His own nature and activity.

How are we, then, to understand the *spatial concepts* embedded in the Creed? In seeking to give an answer we must begin where the Creed itself begins, with God the Father Almighty, Maker of heaven and earth, and of all things visible and invisible—that is, with the transcendence of God over all space and time for they were produced along with His creation. It follows that the relation between God and space is not itself a spatial relation. That is why, as Athanasius argued with the Arians, it is nonsensical to ask of God whether He is without place (χωρὶς τόπου) or whether He is in place (ἐν τόπῳ). Even to put such questions is to presuppose that God can be thought of in a way parallel with ourselves. This also means that the 'came down from heaven' (κατελθόντα ἐκ τῶν οὐρανῶν) which is predicated of the Son is not to be construed in any sense as a journey through space. 'From the heavens' (ἐκ τῶν οὐρανῶν) must be interpreted in accordance with the statements that the Son is 'God from (ἐκ) God, Light from (ἐκ) Light'. The relation between the actuality of the incarnate Son in space and time and the God from whom He came cannot be spatialized. God dwelling in heaven is essentially a theological concept like 'God of God', and no more a spatial concept than God dwelling in Light—even if we could conceive of a heaven of heavens we could not think of this as containing God. It is this biblical way of thinking that is employed here, for God is the transcendent Creator of the whole realm of space and stands in a creative, not a spatial or a temporal relation, to it. The mythological synthesis of God and the cosmos, with its confusion between the presence of God and upper space, is to be found in the anonymous *De mundo* (falsely attributed to Aristotle) that gained currency in the second and third centuries and corrupted proper understanding of Ptolemaic cosmology; but its conception of intervening space between God and man is as far removed from the Nicene theology as anything could be.

On the other hand the statements in the Creed about the descent and becoming man of the Son of God clearly intend to assert His actual presence in space and time and His personal interaction with our physical existence. While the *homoousion* of the Son with the Father expressed the conviction that what He was towards us in His incarnate activity He was inherently, and therefore antecedently and eternally, in Himself, the conjunction of 'came down' with 'for us men and for our salvation' is to be understood as an act of pure condescension on His part and not as an indication of finite imperfection in Him. He was not creaturely or space-conditioned in His own eternal Being, but He humbled Himself to be one with us and to take our finite nature upon Him, all for our sakes. This is what the Nicene theologians called His 'economic condescension'. But the addition to the Creed of the words 'whose Kingdom shall have no end' crowned the conviction that the assumption of our corporeal existence by the Son of God was not '*only* economic', i.e. a temporary episode, for the involvement of the Son of God in our human and creaturely being, even after His resurrection, ascension, and *parousia*, must be maintained without reserve.

This meant that the relation of God to space, not only in respect of His creative and redemptive activity in our physical existence, but in respect of the personal Being of the Incarnate Son, became a very pressing problem for the Early Church. It had to come to terms with the prevailing concepts of space in Greek philosophy and science, and had to forge for itself a more adequate, and in view of the creation and the Incarnation, a more appropriate concept. In so doing Nicene thought developed a *relational conception of space*, in which spatial and conceptual relations were found to be inseparable, but it turned out that the basic problem was more an epistemological than a cosmological one.

The dominating concept in Greek thought was undoubtedly a *receptacle* or a *container* notion of space that went back to the early Pythagoreans and Atomists, and is found throughout the history of Greek philosophy and science, sometimes in a more idealist form and sometimes in a more materialist form, differences that left their mark upon theology as well as astronomy and cosmology. Within this general outlook, however, the prevailing views with which the Fathers had to reckon were those of Plato, Aristotle, and the Stoa.

Plato was prepared to speak of space as a 'receptacle' (ὑποδοχή) but only in a metaphorical sense for he rejected the notion of a receptacle that had the power to set limits to the bodies it contained. Space is simply that *in which* (ἐν ᾧ) events occur, a formless and passive medium (ἐκμαγεῖον) that does not give shape or determination to what is found within it. Plato does not therefore use quantitative or volumetric language when speaking of space, for it is not to be defined in terms of bodies or their extension, but rather in terms of the events that take place within it as the situation it provides for whatever comes into being (ἕδραν δὲ ὅσα ἔχει γένεσιν πᾶσιν). Space and its contents are to be discriminated because there is something permanent and uniform in all change and becoming, but it is not something that corresponds to sensory perception. Because it has no fixed character of its own it is only dimly discernible, but some indication of it may be given in terms of the universal inter-relation of events as the continuum implied in their movement and change (A. E. Taylor).

The fact that Plato brought in the concept of space (χώρα) as a third thing along with archetype (παράδειγμα) and copy (μίμημα) seems to indicate that he thought of space as helping in some way to bridge the separation or chasm (χωρισμός) between the realms of the intelligible and the sensible (the νοητά and the αἰσθητά). Yet space is not the 'receptacle' of the archetypes (παραδείγματα) but of the copies (μιμήματα). Apart from this medium 'in which' they occur the events in the sensible realm would not attain their configuration or therefore their apprehensibility, that is, would not be copies of the archetypes and would not therefore participate in the intelligible forms of the real world. Apart from space we would not be able to penetrate into the rationality lying behind sensible events. Space is necessary for the objective reference of our concepts for it is the medium in which they are grounded in 'that which always is and has no becoming'.

In Plato's epistemology, then, the separation between the sensible and the intelligible realms directs our thought (διάνοια) to an objective ground beyond subjective experience without which it would be inexplicable, but since for this reason the separation between the intelligible forms and sensible things must remain, we may not project the concepts that arise in this world, where they refer to sensible objects, across the separation in order to predicate them of ultimate reality that transcends our world. Rather must we seek to penetrate through to them by following the line of transcendental reference indicated by the rational or 'mathematical' traces (ἴχνη) they exhibit until our mind (διάνοια) rests upon the abiding reality beyond. This is what mathematicians do, Plato claimed, when they make their drawings, which are only like shadows or images in water, for they use them in an attempt to see those things which cannot be discerned except by thought. When this takes place we engage in a movement of thought that Plato spoke of as 'removing the hypotheses' (ἀναιροῦσα τὰς ὑποθέσεις), in which we discard the initial concepts we put forward in order to reach the primary principles to which we are directed by means of them.

This was the problem that faced Plato with regard to the concept of space, or indeed the spatial element in any human concept, when it is applied to the intelligible realm. The separation speaks of a world beyond space and time, and yet a spatial element (χώρα) is involved in the very concept of separation (χωρισμός) itself. Spatial elements have to be used when we speak of what is beyond space and independent of it, yet we cannot project a spatial concept beyond the separation, as if we could speak of there being 'place' over there. What Plato left finally unanswered, however, was the question how far we can discard the very concepts we require in order to reach the objective ground upon which those concepts, in respect of their rationality, must rest, if they are to be true at all. That was to become an acute problem for the Nicene Fathers in their struggle with Arianism, but quite clearly Plato's thought proved a valuable quarry for Christian theologians in their attempt to lay a scientific foundation for theology.

Aristotle misunderstood Plato at two important points—which was not a help to the Fathers: he misconstrued the Platonic separation (χωρισμός) as a local or spatial separation, and mistook the Platonic 'receptacle' or 'matrix' for the original stuff or substrate from which bodies are derived. This was due to his very different approach, from the empirical situations where one body is in fact contained by another and is thus 'in place'. He listed 'space' among the categories and so thought out his conception of it within the substance–accidents scheme of things. As a category, then, space was regarded not only as a fundamental way in which we conceive of things but an actual way in which things exist, and so Aristotle associated space with, and sometimes included it in, the category of quantity (τὸ πόσον or μέγεθος). This led him to develop a predominantly volumetric conception of space, which was reinforced through the attention he devoted to place, or the specific aspect of space (τὸ ποῦ) that concerned him in natural science.

The model that Aristotle used throughout in developing his notion of place or space (ὁ τόπος τι καὶ ἡ χώρα) was the vessel (ἀγγεῖον) into which and out of which things pass and which not only contains them but exercises a certain force or causal activity (ἔχει τινὰ δύναμιν) in relation to them. Place was thus conceived as that which immediately encompasses each body. Aristotle set aside the views according to which the place of a thing is identified with its shape, or its matter, or some sort of extension or interval (διάστημα) between the limits of the container and the thing contained. That left as the only possibility the view that place is the limit or boundary of the containing body at which it is in immediate contact with the contained body. This is a receptacle notion of space in which there is a relation of interdependence between the container and its contents. The normal vessel is a movable place, but place is an immovable vessel. And so Aristotle reached his final definition of place as the innermost unmoved limit of the container (τὸ τοῦ περιέχοντος πέρας ἀκίνητον πρῶτον). That is to say, place is defined as the immobile limit within which a body is contained. Since the containing body is always in immediate limiting contact with what is contained, there can be no void or empty space. There is only magnitude or μέγεθος. While everything within the universe is in place, the universe is not, since it is not encompassed by anything. It may be said to be in place only in the sense that the limit is in the limited. That certainly carried with it the view that space is necessarily limited and finite.

Two problems may be noted here. Aristotle's thought is clearly governed by his demand for a point of absolute rest as the centre of reference for the understanding of change and transition. If everything were in flux we would have no standard by which to gauge anything. That centre of immobility was supplied in Aristotle's cosmology by the centre of the material universe, for although it rotated it did not move forwards or change place. Thus although from his approach to the notion of space through the examination of movement in and out of place, Aristotle appeared to offer a dynamic view of space, he offered instead a rather static concept grounded finally upon relation to a point of absolute rest, which was of course in line with his doctrine of the 'unmoved Mover'. The definition of place as the first unmoved limit of the container involved a further problem, for in equating being in place with a particular volume, it also equated volume with a spatial magnitude. The effect of this predominantly volumetric notion of space was not only to isolate the notion of space from that of time, with all the paradoxes that involves, but to import such a rigidity into the concept of space that it could only be made flexible through a highly artificial disjunction of substance from accidents—the endless difficulties of Western Medieval theology at these points may be taken as sufficient commentary upon these problems.

The Stoics offered a very different approach to space. Like the Pythagoreans and Atomists they were concerned largely with the distinction between space and the void, which they conceived as the distinction between the determinate and the indeterminate, or between 'body' and 'nothingness'—'body' being regarded as 'that which is'. Chrysippus had held that what is 'somatic' is limited or determinate and therefore capable of being thought, but what is 'not somatic' is indeterminate or unlimited and incapable of being thought. Hence, he argued, the notion of space must be thought out not so much from the side of any container as from the side of the body contained. At the same time, the Stoics regarded 'body' as the active principle of the universe, the source of motion and activity, so that space had to be thought out as the function of body in its determinate extension throughout the cosmos. The principle they formulated in this connection was: *σῶμα διὰ σώματος χωρεῖ*, body extends and makes room for itself through body. Space was thus conceived not in terms of the limits of a receptacle but in terms of body as an agency creating room for itself and extending through itself, thus making the cosmos a sphere of operation and place. If this were not so the cosmos could only rest immovable in the unlimited void, for the void by its very nature could not act upon it in any way. The material universe is not held together, then, as Aristotle thought, by an exterior continent or an upper sphere which forces the parts to stay together, but by an interior cohesion or tension (*ἕξις* or *τόνος*) or by an immanent reason (*λόγος*), which manifests itself in the laws of nature as the determinate and rational structure of the universe.

It was this inherent rationality of nature that led the Stoics to speak of the cosmos as a 'rational animal' or to speak of God as 'the soul of the cosmos', for they thought of God, body, and laws of nature, as given together in and with the determinate content of the universe. 'God' and 'body' were in fact different ways of speaking of the physical principle of rationality which could appear equally as mind and as matter. In the nature of the case to think of God as transcendent to the determinate universe would relegate Him to the infinite void beyond space, but on Stoic principles that would be an empty movement of thought. On the other hand, to include God within determinate rationality, and thus within space, meant that the universe could not provide any account of itself or offer any explanation for the rationality of its own structure.

In assessing Stoic doctrine it must be said, on the one hand, that the notion of space as room forged for itself by an active agency was a decided advance over the receptacle notion that prevailed generally in Greek thought. It certainly stood closer to the biblical outlook as seen, for example, in the Old Testament conception of salvation, in which the divine presence is thought of as breaking into man's confined and captive existence, creating space or room in his life for God whom the heaven of heavens cannot contain. On the other hand, by binding the idea of God to a finite universe, conceived as a determinate and limited sphere bounded by the infinite void yet having its centre in the earth, the Stoics not only failed to reach, or lost, any understanding of the transcendence of God, but made it too easy for the popular and certainly the pagan mind to confound God with nature, theology with cosmology. This is in fact what we see in the *De mundo* of the second century which was evidently a product of debased Stoic thought.

We must now ask how the Nicene theology reacted to these concepts of space, and how far it made use of them in thinking out the evangelical message of the coming and presence of

God in space and time in Jesus Christ. As far as I am aware the Aristotelian definition of space found no place at all in the Nicene theology. Its rejection meant also the rejection of a strict receptacle view of space and the cataphatic conceptualism that went along with it—it was only with John of Damascus that this idea gained modified recognition in the East, although it played a considerable role in the West, especially after the twelfth century. Certain Platonic and Stoic notions, however, proved more helpful to the Christian Church; but the Nicene Fathers were forced to think out the whole question from primary theological grounds, especially in relation to the doctrines of creation and incarnation.

The doctrine of creation out of nothing, which very quickly came to the forefront of Christian thought, asserted the absolute priority of God over all time and space, for the latter arise only in and with created existence and must be conceived as relations within the created order. They are orderly functions of contingent events within it. Time is in creation, creation is not in time. Since God is the transcendent source of all that is, beyond Himself, it may even be said of Him that He does not participate in being, for all else participates in Him for its existence. God Himself, then, cannot be conceived as existing in a temporal or spatial relation to the universe. If traditional Greek concepts are to be used, it must be said that God is not contained by anything but rather that He contains the entire universe, not in the manner of a bodily container, but by His *power*. Everything that is 'body', even in the Stoic sense, whether visible or invisible, is contained by that divine power, but since it comprehends and encompasses everything there can be no infinite void, for in His own transcendent way God is everywhere and in all things. It follows from this that space and time, and indeed all the structured relations within the universe, have to be understood *dynamically*, through reference to the creative and all-embracing power and activity of God.

It was Origen who was the first to discern the philosophical significance of this reversal of Aristotelian and Stoic concepts, in establishing the connection between the transcendence of God and the rationality of nature, thus delivering the universe from being shut up in the futility of being unable to offer any explanation of its own rationality. Origen accepted the Stoic principle that comprehension and limitation go together, for what is not limited or determinate (πεπερασμένον) is incomprehensible. That is why we speak of God as incomprehensible, for He is immeasurable and far transcends all our thoughts about Him. But Origen reversed the operation of the Stoic principle by insisting that it is the fact that *God comprehends all things* that limits them, giving them beginning and end, thus making them determinate and comprehensible. He did not hesitate to point out the implication that God is limited by His own rationality in knowing Himself, but self-limited also in the limitation of the creation through its subjection to His comprehension, oversight and providence. The rationality of the universe is thus grounded in the creative comprehension of it by God, but it is this 'comprehending' or 'containing' (περιέχειν) that provides the rational structure within which God is revealed and through which theological statements are given their objective and cognitive reference to God. It is not difficult to see the link between this and Plato's doctrine of space as the medium in which we cognize the correlation of events and through which we trace their reference objectively to the abidingly real. But space has become here an epistemological as well as a cosmological principle.

Origen also had to face the question that was posed for Plato in view of the separation or χωρισμός between the sensible and the intelligible worlds. If we may not project our sense-concepts beyond the separation, do we have to discard them altogether in order to think the real? This was made more acute for Origen by his immense stress on the transcendence of God. On the one hand, the separation between God and the world was made so sharp that it became necessary to think away all the forms and shapes we derive from our contemplation of creaturely existence in order to think worthily of Him, but on the other hand, the fact that God confers upon this world rational and mathematical character by maintaining it as the object of His knowledge and power meant that we cannot entirely discard spatial and temporal concepts when speaking about the other world, without lapsing into incomprehensibility and irrationality. Origen looked for the solution of these questions in the Incarnation of the Word or Son of God through whom all things have been created. By coming into this world to be one with us where we are He has established in Himself a real and rational relation between men on earth and God the Father in heaven—i.e. as Origen did not hesitate to say, between 'earthly place' and 'heavenly (or 'superheavenly') place'. In explaining this Origen had recourse to the Stoic idea that space arises when 'body', or the active principle, makes room for itself and that the concept of space or place must be formed in accordance with the nature of the occupying agent. Of course, it is not 'body' but the Creator Word of God who is the active principle or agent. The incarnation means that He by whom all things are comprehended and contained by assuming a body made room for Himself in our physical existence, yet without being contained, confined or circumscribed in place as in a vessel. He was wholly present in the body and yet wholly present everywhere, for He became man without ceasing to be God. He occupied a definite place on earth and in history, yet without leaving His position or seat (*ἕδρα*—it is Plato's term that is used) in relation to the universe as a whole. Thus the Aristotelian concept of place or space was sprung wide open. Clearly Origen began to form a relational notion of space, but he ran into considerable difficulties through his cosmological speculations and through applying the idea of limiting comprehension to Christ's knowledge of the Father. He did, however, lay the basis for the Nicene conception of space which was developed and defended with particular force by Athanasius.

The essential key was found in the relation of the *homoousion* to the *creation*, that is, in the fact that the Lord Jesus Christ, who shares with us our creaturely existence in this world and is of one substance with the Father, is He through whom all things, including space and time, came to be. That is to say, Nicene theology rejected an approach to the questions of space and time from any point of absolute rest, but adopted an exclusive approach from the living and active Self-Word (*Αὐτολόγος*) through whom all things, visible and invisible, were created out of nothing, who leaves nothing void of Himself, and who orders and holds the entire universe together by binding it into such a relation to God that it is preserved from breaking up into nothingness or dropping out of existence, while at the same time imparting to it light and rationality. He it is who by Incarnation has come into our spatial realm (*εἰς τὴν ἡμετέραν χώραν*) although He was not far off before. He is fully present with us in space and time and yet remains present with the Father. He was certainly 'outside' all things in respect of His essence, but 'in' all things and events in respect of His power. Yet He did not activate the body He assumed from us in such a way as to cease to deploy Himself actively throughout the universe in all its dimensions. Since He shared with us our physical space, while remaining what He ever was, the spatial ingredient in the concept of the Incarnation must be interpreted from the side of His active and controlling occupation of bodily existence and place.

Space is here a predicate of the Occupant, is determined by His agency, and is to be understood in accordance with His nature. He cannot, therefore, have the same space-relation (χώρα) with the Father as we creatures have, otherwise He would be quite incapable (ἀχώρητος) of God.

Unlike Plato and Origen, Athanasius never operated with the concept of separation (χωρισμός) between the sensible world (κόσμος αἰσθητός) and the intelligible world (κόσμος νοητός), for the linking together of incarnation and creation in the manner of Nicaea made that impossible. The two worlds came to be understood not as entirely separated, nor as only tangentially related to each other, but as actually intersecting in Jesus Christ. Thus, as Athanasius argued, through His relation with the Father and His relation with us, the incarnate Son of God fulfils the part of a Mediator (μεσίτης) even in regard to space-relations between man and God, for mere creatures are unable to make room (χωρεῖν) for God in their natures, far less are they able to endure the Creator in their created beings. This bridge is supplied in the Incarnation but the relation between the incarnate Son and the Father cannot be thought out in terms of a receptacle (ἀγγεῖον) notion of space, for the application of such a concept to the *kenosis* can only lead to a false kenoticism which does not do justice to the 'fullness' and 'perfection' of either the Father or the Son, since it fails to think of them in accordance with their natures. The inter-relations of the Father and the Son must be thought out in terms of 'abiding' and 'dwelling' in which each wholly rests in the other. This is the doctrine of the *perichoresis* (περιχώρησις) in which we are to think of the whole being of the Son as proper to the Father's essence, as God from God, Light from Light. Creaturely realities are such that they can be divided up in separated places (ἐν μεμερισμένοις τόποις) but this is impossible with the uncreated source of all Being, with the Father,

Son and Holy Spirit who wholly dwell in each other and who each have room fully for the others in the one God. Now when the Son, who abides in the Father in that way, became incarnate, He became for us the 'place' where the Father is to be known and believed, for He is the τόπος or *locus* where God is to be found. But 'place' is here *stretched* beyond its ordinary use, and must be interpreted *elastically* in accordance with the nature of God and His activity in revelation and redemption through the Incarnation as well as in accordance with the nature of the creaturely human being in which He became incarnate for our sakes. This forces theology into the construction of a sort of *topological* language in order to express the dispositional and dynamic inter-connection between *topos* and *topos* or place and place. The fact that this requires a differential use of concepts in which the ordinary and natural concept of place or space had to be adapted and changed, did not trouble Athanasius, for that, he held, is what must happen when we use terms and concepts rightly in accordance with the nature of the subjects they are employed to denote.

Christ is 'in' us through sharing our bodily existence, but He is also 'in' the Father through His oneness with Him, but how are we to work out the relation between these two 'in's? Athanasius offered an analogical account of the relationship. For this purpose he constructed and employed the concept of *paradeigma* (παράδειγμα) which is not to be translated as 'model' or 'representation', far less as 'archetype', but may be translated as 'pointer', for it is essentially an operational term in which some image, idea or relation is taken from our this-worldly experience to point beyond itself to what is quite new and so to help us get some kind of grasp upon it. Yet it can succeed in this function only so far as we understand the *paradeigma* itself in the light of the reality (κατά τι θεωρούμενον) it serves. He used *paradeigma*, then, to refer to the kind of concept which under the impact of the divine revelation is made to point beyond its creaturely form and content to the intended reality (πρὸς ἄλλο τι βλέποντα), without transgressing the distance (διάστημα) or rubbing out the difference (διαφορά) between them. It is a cognitive instrument enabling us to get some mental hold on the reality revealed, not one by means of which we capture it through our conceiving of it. It fulfils its function while making evident its inadequacy by pointing intelligibly to what is really apprehensible although ultimately beyond the limit and control of our comprehension.

In the nature of the case, the *paradeigmata* (παραδείγματα) that we employ in theology are not those we choose, but those that are thrust upon us through divine revelation, and which have their ultimate ground, correction and validity in the relation between the Father and the Incarnate Son, and the Incarnate Son and the Father. That is the relation that bridges the separation (χωρισμός) between God and man and supplies the epistemological basis for all theological concepts, and therefore for our understanding of the relation between their creaturely content and the reality of God Himself. It is in Christ that the objective reality of God is intelligibly linked with creaturely and physical forms of thought, so that the latter may be adapted and given an orientation enabling them to direct our minds to what God really makes known of Himself, although in view of His infinite nature they will not be able to seize hold of Him as He is in Himself.

It was by using *paradeigma* in this way that Athanasius sought to relate the being and activity of the Son of God to bodily place (τόπος) when He entered into our human space (χώρα) and became man, without leaving God's 'place' and without leaving the universe devoid of His presence and rule. Since space is regarded here from a central point in the creative and redemptive activity of God in Christ, the concepts of space as infinite receptacle or infinite substance, or as extension conceived as the essence of matter, or as a mere necessity of our human apprehension, and certainly the concept of space in terms of the ultimate immobile limit of the container independent of time, all fall away, and instead there emerges a concept of space in terms of the ontological and dynamic relations between God and the physical universe established in creation and incarnation. Space is here a *differential* concept that is essentially *open-ended*, for it is defined in accordance with the interaction between God and man, eternal and contingent happening. It is treated as a sort of coordinate system (to use a later expression) between two horizontal dimensions, space and time, and one vertical dimension, relation to God. In this kind of coordination, space and time are given a sort of transworldly aspect in which they are open to the transcendent ground of the order they bear within nature. This means that the concept of space which we use in the Nicene Creed is one that is relatively closed, so to speak, on our side where it has to do with physical existence, but is one which is infinitely open on God's side. This is why frequently when Byzantine art sought to express this *ikonically* it deliberately reversed the natural perspective of the dais upon which Christ was represented. The Son of God become man could not be presented as one who had become so confined in the limits of the body that the universe was left empty of His government. He could not be represented, therefore, as captured by lines which when produced upwards met at some point in finite space, but only between lines which even when produced to infinity could never meet, for they reached out on either side into the absolute openness and eternity of the transcendent God.

This is surely characteristic of the concepts which we use in Christian theology. Of course the terms and concepts we use are human and creaturely, with human and creaturely content, for they belong to our existence in this world and partake of the limits of nature. If they are to be used to speak of God they must be stretched and extended beyond the range of the phenomenal world for which they were formed in the first place—otherwise they can serve only to exclude everything except purely natural knowledge. We are aware of something like this in the advance of natural science where we must be prepared to do violence to our ordinary forms of thought and speech if we are to apprehend what is genuinely new. We have to devise new languages and step up to higher levels of thought in order to push knowledge beyond the limits of ordinary experience, yet all this remains within the limits of nature, for we ourselves belong to nature and are unable to rise above it. We cannot of ourselves transcend the necessities of the finite apprehension. Scientific violence of this kind is a more stringent requirement in theology, for a considerable shift in the meaning and reference of our ordinary terms and conceptions is necessary if they are really to indicate God Himself. But if that is to happen, they must be opened up from above, as it were, for anything we do to push our words and concepts beyond the boundary of creaturely being can only take a mythological form, that is, by way of projecting the creaturely content of our statements as such on to God. To be more precise, if human forms of thought and speech are to have a transcendental reference to what is really beyond them, it must be given them by God Himself. That is why theological statements have an orientation from beyond themselves in the coming of the Word of God and through assimilation to that Word are taken up to a higher level of understanding where, as Edward Schillebeeckx has put it, they partake of a new dimension which they do not possess of their own accord.

On the other hand, theological statements must retain some genuine connection with our plain, straightforward language, for if the concepts they embody are completely detached from those that are found in our ordinary knowledge, the statements could no longer mean anything to us, far less convey anything to others. This is of course characteristic also of natural science, for the highly abstract denotations which we have to develop in order to lift the range and level of our knowledge cannot be cut adrift from the physical language based on our old concepts of space and time, for it is only through them that the applicability of scientific concepts and terms, and their truth, can be shown. Thus, while scientific concepts need to be extended in order to lay hold of what is new, they cannot be allowed to take off into an arbitrary world of their own, but must be held in a structure of levels in which the lower levels are opened to the higher levels and the higher are controlled through coordination with the lower. In the nature of the case the connection between the levels can be made only *after* new knowledge has been gained and new concepts have been formed.

It cannot be otherwise in a disciplined theology. The scientific function of theological statements is to offer a rational account of knowledge beyond the limits of mere this-worldly experience through the use of acknowledged concepts taken from this world, and so to help our minds to lay hold upon it even though it is more than we can grasp within the limits of these concepts. Theological statements properly made are thus by way of being operational statements directing us towards what is new and beyond but which cannot be wholly indicated or explained in terms of the old. This was well expressed by Hilary: 'There can be no comparison between God and earthly things, but the weakness of our understanding forces us to seek certain images from a lower level to serve as pointers to things of a higher level (*species quasdam ex inferioribus, tanquam superiorum indices*). . . . Hence every comparison is to be regarded as helpful to men rather than as suited to God (*homini potius utilis, quam Deo apta*) since it suggests rather than exhausts the meaning we seek (*intelligentiam magis significet, quam expleat*).'[1] Hence while theological concepts *must retain their creaturely content*, we cannot claim to lay hold of the divine reality by means of them. Nevertheless they must be employed in the service of what is *given* in our knowledge of God, in an act of objective intention in which their creaturely content is not ascribed to God as such, but becomes the medium of transcendental reference to Him. Theological statements operate, then, with essentially *open concepts*—concepts that are relatively closed on our side of their reference through their connection with the space-time structures of our world, but which on God's side are wide open to the infinite objectivity and inexhaustible intelligibility of the divine Being.

[1] *De Trinitate*, 1.19.

CHAPTER 2

THE PROBLEM OF SPATIAL CONCEPTS IN REFORMATION AND MODERN THEOLOGY

In modern as in ancient thought some form of the receptacle notion of space seems to have been predominant. Only in comparatively recent science have we departed from it, but this step continues to involve us in a struggle with age-old habits of mind for, as A. C. Ewing has pointed out, we will persist in speaking of things as *in* space or *in* time. There may well be at work here a substitute symbolism arising out of post-natal desire for the security of being at home in a container and anxiety at being thrust out into the open world where we are not enveloped by finite constants but are exposed to limitless and incomprehensible immensities. Be that as it may, throughout history the receptacle notion of space has proved very troublesome for theology, but perhaps never more than in modern times when space came to be thought of as a container independent of what takes place in it and regarded as an inertial system exercising an absolute role in the whole causal structure of classical physics.[1] In spite of the new physics, the problems to which this gave rise are still with us, and not a little of the confusion in current theology may be put down to them.

[1] Cf. Einstein's interpretation of this in his introduction to Max Jammer, *Concepts of Space*, p. xv.

The Church Fathers had to face similar problems posed for them from the side of classical culture. On the whole Greek science and philosophy had come to concentrate upon the determinate and finite, for that alone was conceivable, and so operated with a notion of space as delimited place defined in terms of a containing vessel. If there were an actual (as distinct from a mathematical) infinite it could only be the limitless, empty void beyond the cosmos, but although some such notion was advocated by Pythagoreans and Atomists alike, Platonic, Aristotelian and Stoic thought deemed it unthinkable and unintelligible. Thus the finite, the intelligible and the limits of the container were all bound up together in their thought. This proved to be quite impossible for the Church Fathers, since it implied that if God is intelligible He must be finite. For Christian theology, however, God is the Self-existent, infinite and eternal, the Maker of heaven and earth, so that all things, visible and invisible, must be understood through reference to His creative and all-embracing wisdom and power. Hence instead of thinking of God in accordance with the determinate features of the finite cosmos the theologians of the early Church thought of Him as the transcendent Source of all rationality who, by maintaining the universe as the object of His creative knowledge and power, structures and limits it, making it determinate and comprehensible. It was within this context that they approached the questions of space and time.

The doctrine of creation out of nothing meant that God does not stand in a spatial or temporal relation to the universe but that spatial and temporal relations are produced through His creation of the universe and maintained through His interaction with what He has made. Moreover, since all creation is comprehended by God and endowed with rationality in this way, space and time are to be conceived not only as relations arising in and with created existence but as bearers of its immanent order. This is not contradicted by the doctrine of the Incarnation according to which God Himself has entered into our world in Jesus Christ through the assumption of a physical body in space and time, but is confirmed and illuminated, for, while it does not discount the absolute priority of God over all space and time, it asserts the reality of space and time for God in His relations with us and binds us to space and time in all our relations with Him. Thus the miraculous activity of God in the Incarnation is not to be thought of as an intrusion into the creation or as an abrogation of its space-time structure, but as the chosen form of God's interaction with nature in which He establishes an intimate relation between creaturely human being and Himself. Here space and time provide the rational medium within which God makes Himself present and known to us, and our knowledge of Him may be grounded objectively in God's own transcendent rationality.

Thus it came about that in seeking to articulate its understanding of God's activity in creation and Incarnation, Patristic theology rejected a notion of space as that which receives and contains material bodies, and developed instead a notion of space as the seat of relations or the place of meeting and activity in the interaction between God and the world. It was brought to its sharpest focus in Jesus Christ as the place where God has made room for Himself in the midst of our human existence and as the place where man on earth and in history may meet and have communion with the heavenly Father. But because this was a conception in which the theologians of the Church were forced to ask questions in opposite directions, developing their thought in accordance with the nature of the Creator who transcends all space and time and in accordance with the nature of the creature subject to space and time, it became essentially a *differential and open concept of space* sharply opposed to the Aristotelian idea of space or place as the immobile limit of the containing body. Although this was a distinctively theological understanding of space that was not tied down to any particular cosmology, it was capable of being translated into purely physical terms, and did in fact leave its mark on the development of ancient physics, as we can trace in the thought of Damascius and Philoponus.

When we turn to the thought of the West it is a different story that must be told. From early times Latin Christianity had assimilated the idea of the receptacle into its theology where it affected deeply the form in which the doctrines of the Church, Sacraments and Orders were developed. Supernatural grace was widely thought of as contained in ecclesiastical vessels and capable of being handed on in space and time by means of them. This is one of the significant points where we find the Latin language moulding Christian experience to specific forms of thought. But it was when Aristotelian philosophy and science became ascendant in the twelfth and thirteenth centuries that the receptacle notion of space was consolidated in the whole structure of Medieval theology. The fundamental approach to the question of space was altered through the Aristotelian schema of substance and accidents, so that it came to be thought out not primarily as in the East from a consideration of creation and Incarnation but from a consideration of the real presence of Christ in the Mass on the one hand and from a consideration of cosmology on the other hand. It was inevitable that these two approaches should be related much as the Aristotelian definition of place as the *terminus continentis immobilis primus* (to give it its scholastic form) was related to a centre of absolute rest in the earth and in the unmoved Mover behind it. Moreover this led the Medievals to think of the presence of God almost entirely in a *spatial* manner, *apart from time,* which helps to explain the inadequate place they gave in their thought to a dynamic relation between God and historical existence and to the temporal relevance of the Last Things. In 1277 a great attempt was made in the University of Paris to reject the Aristotelian and Averroist theory that since motion presupposes an immovable body there must be a point of absolute rest, and so to put back into the centre of the theological and cosmological picture a Prime Mover exercising absolute power, that is, a doctrine of God who is able to move the whole universe through space. But before very long Aristotelian thought again prevailed, while the notion of God's absolute power was used in order to extricate theology from the difficulties and contradictions resulting from the acceptance of Aristotle's definition of place.

Medieval theology took over from Patristic theology the conception of the rationality of nature and developed it in such a remarkable way that the whole of modern science is indebted to it, but when this was allied to its predominant movement of thought *from the world to God*, it meant that Medieval theology systematically developed a 'natural theology' *detached from all actual relations with God* which, as a *praeambula fidei*, then became the framework within which what was called 'revealed theology', was inevitably interpreted. It is in this context that we can see the problem created by the Aristotelian notion of space worked out apart altogether from the interaction of God with nature. Since according to Aristotle there is a relation of interdependence between the containing vessel and what it contains, this receptacle notion of space cannot but exercise conceptual control over whatever is conceived by means of it. This is, of course, the 'cataphatic conceptualism' of Latin theology, whether in its realist or nominalist forms, against which Medieval and Modern Greek theology have insistently protested.

So far as Christology was concerned classical Medieval theology in the West began by following the lead of John of Damascus who was apparently the first to introduce Aristotle's definition of place or space into Eastern theology, and so insisted that in Christ God became man without ceasing to be God, entering into physical space without being circumscribed by it. But Latin theology failed to follow John of Damascus in the way in which he worked this out in terms of the divine activity (ἐνέργεια) and what he called 'mental space' (τόπος νοητός) which enabled him to escape from a cataphatic rigidity of thought. A move in this direction was hardly possible for Latin theology so long as it remained within the orbit of Aristotelian thought that categorized space as a quantity (μέγεθος) and tied understanding of it to the substance-accidents, subject-predicate way of thought, but it did become possible during the Renaissance when this way of thought began to break up and a new physics struggled to emerge.

Late Medieval thinking about space, however, was activated for the most part by problems that arose out of the idea of the *real presence*. How can the body of Christ be contained in the host, and how can it be in many hosts at the same time? The doctrine of transubstantiation had attempted to get round the rigidity of the Aristotelian concept of the container while taking seriously the real presence of the body of Christ. But this was achieved only at the cost of a highly artificial separation between substance and accidents and was questioned by Occamists, empirical Aristotelians and Nominalists, on the ground that the container is not independent of what it contains—space is not different from a spatially perceived object. Aristotle himself held that space is an accident of substance. But they on their part were faced with the problem of how to maintain the real presence of Christ in His body and blood without any commensurate relation between them and the quantitative features of the containing vessels, the bread and wine, which they held must remain even when their substance is converted into the body and blood of Christ. They were forced to invoke the absolute power of God to enable the whole Christ to be in the whole host and in each part of it, without His body being confined or circumscribed in a place in the sacrament of the altar, while nevertheless asserting that the material substance through its intrinsic parts is present in the place in which it is circumscribed. Thus they distinguished a special kind of presence applicable to Christ according to which His body could be regarded as present through something else that is circumscribed in its place. Moreover, since this is conceived in a non-quantitative or non-metrical way, it can be conceived only as an extensionless or mathematical point. This view was often found at the end of the fifteenth and at the beginning of the sixteenth century and was to influence Martin Luther. But all the variations in the later Middle Ages among Thomist and Occamist notions retained the receptacle or container concept of space. Obviously to interpret that strictly in the quantitative or volumetric manner of Aristotle is to introduce impossible paradoxes and absurdities into theology, but even when these can be stripped away the fundamental difficulty remains in the correlation of receptacle and space. When therefore the Renaissance physicists rejected the Aristotelian concept of space only to return to that of the Atomists in which space was distinguished as a receptacle independent of what goes on in it, and went on to advocate the priority of space over matter, they began to set the stage for centuries of difficulty.

There was, however, another but much smaller stream of thought running out of the Middle Ages that went back to Patristic and Platonic sources, and was not without its influence at the Reformation especially in Reformed and Anglican theology. Although Grosseteste made some use of Aristotelian terminology he declined to develop the idea that space or place is to be understood in terms of the containing body, and like Plato thought of it as separate from what goes on in it but as having no being or dimensional character of its own. It arises whenever bodies are set in motion and serves to distinguish one thing from another. In line with this, Pseudo-Grosseteste held body to be a pure relational continuum and he seems to have regarded place as some sort of location in the inter-relation of dimensional bodies.[1] This was further elaborated in an exciting theory of light as the first corporeal form and the first principle of motion and therefore as the ultimate basis of extension in space.[2] Duns Scotus also developed a more dynamic and relational concept of place through reflecting upon creation and incarnation, i.e. in the light of the fact that all creation is contingent upon the freedom of God's creative Will and is related to the active and creative ideas which God freely and rationally produces along with the created realities themselves, and in the light of the fact that since Christ became incarnate in such a way that He did not leave off His operations in the universe, He can be in heaven and in an infinite number of places at the same time. Thus Duns Scotus approached the question of space not from a point of immobility in the universe or immutability in God but from a centre in God's active relation to the world. In line with his conception of existence and individuation expounded under the term *haecceitas*, he laid the emphasis upon the distinctive nature and mode of presence, that is, the individuating *hic esse* of bodies in motion. There is a distinctive *hereness* corresponding to distinctive *identity*. Hence he detached the notion of

[1] See C. K. McKeon, *A Study of the Summa Philosophiae of Pseudo-Grosseteste*, p. 114.

[2] This is briefly discussed by M. Jammer, op. cit., p. 36 f., who also links it up with the Neo-Platonic tradition.

space from matter and thought of it in terms of location rather than containing place, or position rather than volume. This allowed him in the sacrament of the Mass to hold a view of the real presence as the active relation of the body of Christ to the host, which cut out the need for any theory of transubstantiation. It is more along this line that we are to look for the Medieval antecedents to the thought of John Calvin, although he was certainly more deeply affected by Patristic conceptions.

In turning now to the Reformation we must focus our discussion mainly upon Lutheran thought since it was largely owing to its retention of the receptacle notion of space in the doctrines of the real presence and the incarnation that an alliance was made possible between Protestant theology and the new physics of the seventeenth and eighteenth centuries, which also operated with a receptacle notion of space. In these circumstances it was inevitable that the Medieval problems should be carried over into modern theology, but it is in Lutheran thought that they have appeared in their most striking form. The receptacle notion of space was immensely important for Luther for it was his way of asserting the reality and actuality of the Son of God in our human and earthly existence, and so he concentrated his thought with a furious intensity upon the fact that the whole Son and Word of God is contained in the infant of Bethlehem and communicated to us in the sacrament of the Lord's Supper. Reformed and Anglican theology, as we can see from an examination of the teaching of Calvin and Hooker, stood much closer to the thought-forms of classical Patristic theology, although it was set out consciously *vis-à-vis* the problems that arose out of debates with Lutherans regarding the self-emptying of Christ and the ubiquity of His body. It will be sufficient for our purpose to refer only to three points of difference between the

Lutherans and the Reformed (who were here at one with the Anglicans), viz. the so called '*extra Calvinisticum*', the 'location' of the body of Christ in heaven, and the Eucharistic *parousia*.

When the Reformed theologians spoke of the Son of God in His Incarnation as having descended from heaven without leaving heaven, or as living and acting on earth without abandoning His government of the universe, they were deliberately following the early Church in the rejection of any receptacle notion of space that demanded the enclosure or confinement of the Son of God in a human body, and indeed in the rejection of any local or spatial connection between the divine and human natures of Christ. But the Lutherans operating with a receptacle or container notion of space could only interpret this to mean that in the Incarnation something of the Son or Word of God was left *outside*, *extra*—which they dubbed 'the Calvinist extra'. Likewise when the Reformed theologians spoke of the Ascension as a movement above the heaven of heavens, beyond the created universe, to the right hand of the Father, transcending all our conceptions of space and time, and yet spoke of the location of Christ's body in heaven, they were again rejecting a closed receptacle notion of space while insisting that in His Ascension the body of Christ did not lose dimensional character or reality as human body. As the Incarnation meant the entry of the Son into space and time without the loss of God's transcendence over space and time, so the Ascension meant the transcendence of the Son over space and time without the loss of His incarnational involvement in space and time. Thus when they spoke of the Ascension of Christ from place to place they were adopting the open and differential concept of space developed by the Church Fathers, interpreting 'place' differently in accordance with the nature and activity of God on the one hand and in accordance with the nature and activity of man on the other hand. The Lutherans, however, were unable to follow this and on their own presuppositions could only read the language about the body of Christ in heaven to mean that it was confined there as in a container in the way that they conceived of the Incarnation. Again, when the Reformed theologians expounded the real presence of Christ in the Eucharist but because of the Ascension distinguished it from the real presence of Christ at the Last Day, they were rejecting a receptacle notion of space which cut off spatial from temporal relation, and insisting on understanding the Eucharistic *parousia* as the active self-presentation of Christ to us in space and time that reaches out towards the consummation in the final *parousia*. The Lutherans were unwilling to consider this, since from their point of view it appeared to detract from the full actuality of the real presence of Christ in the Sacrament, yet they could maintain their view only at the expense of making the real presence timeless, i.e. a purely spatial presence unconditioned by time.

How then did Lutheran theology deal with the problems of space? Luther took over from the Occamists a distinctive notion of 'presence' (*praesentia*) which they developed in their application of Aristotle's concept of place as a containing vessel, according to which a thing may be in a place without being able to be measured according to the space of the place, but without resorting to transubstantiation. However, instead of lapsing into the subtleties of the Nominalists to explain this, he sought to understand it by reference to the Incarnation of the Son and the absolute power of God. Taking a cue from Occam he saw in the Patristic doctrine of the *communicatio idiomatum* a means of justifying the literal understanding of *hoc est corpus meum*. But in doing this he transferred the concept of *perichoresis* from the Trinity to the hypostatic union and so developed a notion of the coinherence of the divine and human natures that more than risked the deification of Christ's humanity. On the other hand, Luther insisted that everything must be seen in the light of the mighty, living, active power of God. Sometimes he could speak of God's ability to reduce the whole universe to a grain of sand which being so little could be spanned by the humanity of Christ, and argued that there should be no difficulty, therefore, in understanding how God can make the body of Christ present in the sacramental elements. At other times he could speak of 'the right hand of the Father' as God's omnipotent power, so that to be at God's right hand or to be in the presence of God is to be connected up with His omnipotence and not to be in a place. All this imported a dynamism into Luther's thought, so that the spatial ingredient in the doctrine of Christ or of the Eucharist was understood not in a static manner from a point of absolute rest, but from a centre in God's creative and almighty activity. This meant that the presence of God in Christ, in His Word or in the Sacrament, is entirely contingent upon His Will to be present. Luther drew a distinction, therefore, between the presence of God in Himself and His presence *for you*, corresponding to his epistemological distinction between God in His bare reality in Himself which is hidden from us (*Deus absconditus*) and God in His reality for us as He is revealed to us (*Deus manifestus*). Because God transcends all things He is free and unbound wherever He is, without any kind of spatial tie. And so Luther could say: 'Even God is not where He is, even if He is everywhere or somewhere.' But God is present to us in so far as He *wills* to be present *for us*, binding Himself to us for our good and directing us to His humanity in Christ as to a place—which He does through the *Word* that He addresses to us.

All this is to be understood, as Luther repeatedly insisted, in the light of the distinction which, following Augustine, he drew between the two kingdoms, the eternal and the temporal, the invisible and the visible, the kingdom of faith and the kingdom of sight. They are so sharply divided that they meet only tangentially, that is (and this is one of Luther's favourite ideas), only at a *mathematical point.* That is the language he used to speak of the place where God wills to be present for us. Just as Luther understood eternity as a 'total simultaneity' in which all the succession of time is present to the eternal as a timeless moment, so he understood the presence of God to us as one in which all spatial relations are reduced to a mathematical point in God's will to be present for us in His Word. This idea of God's *instanteity* in His Word was to have far-reaching implications.

Luther's treatment of the problem of space is thus essentially dialectical. On the one hand, he operated with a strictly receptacle idea of space in his doctrine of the real presence. This enabled him to lay stress upon the *hoc est corpus meum* which became for him the ontological nail that held the two kingdoms together, and made possible real participation of one in the other. He knew that if ever there were to be substituted for the *hoc est* a *hoc significat*, the two kingdoms would fly apart and all that would be left would be an unbridgeable dualism with only a paradoxical and vague symbolic relation between them. Thus the *hoc est corpus meum* gave to the receptacle notion of space a fateful necessity in the history of Lutheran theology. On the other hand, Luther brought back the biblical conception of the living and active God which had tended to drop out of sight in the Medieval world, but which now reintroduced a dynamic relation between God and the world that eventually helped to destroy the static conception of space that prevailed in the Middle Ages. But, taken along with the way in which he distinguished between the two kingdoms, this dynamism had the effect of interpreting the presence of God as the moment of the divine Will 'for you' given in the Word of God, and thus of reducing relations with God to mathematical points of timeless encounter. Without doubt an idea of a spatial presence that ignores time, such as we can find in Medieval thought, tends to lapse into magic, but it is no answer to that to reduce spatial relation to a mathematical point, for that is still to exclude time from our relations with God. If we posit any kind of spatial relation without extension in time we make it impossible to discern any difference between the real presence of Christ in the days of His flesh, in the Eucharist, and at the Last Day; but that is equivalent to making the historical foundation of faith irrelevant.

It was thus that Luther set the pattern for the development of German Protestant theology. It is only in recent times that the second strand in his thought has come powerfully to the front, but in the history of classical Lutheran theology it is the first strand that became dominant. This was formulated by Luther's immediate followers under the principle *finitum capax infiniti* which had the effect of consolidating the receptacle notion in the epistemological structure of Lutheran thought. This was precisely the point discussed by the Nicene theologians when they rejected the idea that the human creature is able to make room (χωρεῖν) for God in its nature and is able to grasp or comprehend (χωρεῖν) the infinite, and so insisted that the receptacle (ἀγγεῖον) notion cannot be used to interpret the Incarnation as the emptying of the Son of God into a human vessel, for the human creature is not χωρητὸς τοῦ ἀπείρου or, to put it back into Latin, *capax infiniti*. The difficulties that the Lutheran principle involved soon came to the fore at two points, in their theories of *kenosis* and the ubiquity of the body of Christ. The receptacle notion of space meant that they had to think of the Incarnation as the self-emptying of Christ into the receptacle of a human body (unless they referred the *kenosis* only to the Incarnate Son), but at this point they were faced with exactly the same problems that the Medieval theologians had come up against in the real presence of Christ in the host. It is interesting to note that not only the Medieval problems but the Medieval attempts to solve them now found their counterparts in the rival schools of Giessen and Tübingen, viz., in the Lutheran attempts to think out the real presence of God in Christ either by the renunciation of certain of the properties of the divine nature distinguished from others, or, if they could not be separated in this way, by their concealment and restriction in order to preserve the form of the human receptacle. But the same difficulty cropped up when they focused attention upon the nature of the creaturely receptacle, which was then met by adding to the *finitum capax infiniti, non per se sed per infinitum.*[1] In other words the receptacle had to be enlarged in order to make it receive the divine nature within its dimensions, and so it was held that the Son of God communicated to the humanity of Christ an infinite capacity enabling it to be filled with the divine fullness. In regard to the real presence in the Eucharist this took the form of the ubiquity of the body of Christ.

There can be no doubt that the theological intention of the Lutherans was to assert as strongly as possible the actual condescension of God to be one with us in our existence in space and time, and yet to preserve the essential unity of the divine and human natures in the one Person of Christ, thus making Christ the centre and norm of all our ontological and epistemological relations with God, but quite evidently the non-theological assumptions with which they operated in the receptacle concept of space forced them to take up a position

[1] This is the way Bonhoeffer expressed it, *Christology*, p. 96. For a modern Lutheran use of the receptacle idea see M. J. Heinecken, *Marburg Revisited* (ed. by P. C. Empie & J. I. McCord), pp. 81 ff.

which imperilled the reality of Christ's human nature, so that under the growing pressure of empirical science within the actualities of secular extension, some form of radical dualism was bound to emerge. But before we take up that story we must turn to see what became of the concept of space in the new scientific culture.

Out of the Renaissance there had come the notion of space as the infinite receptacle of all things, which was held along with the Platonic idea, revived through the Florentine Academy, that space is the medium in which sensible events partake of mathematical character and therefore become intelligible. But this could be maintained over against the Aristotelian and Stoic view (which dominated ancient science) that the world is intelligible only because it is finite, if space were somehow aligned with the infinity of God. That is what began to happen after the Reformation when the predominant movement of thought in the Middle Ages gave way under theological pressure from the doctrine of *grace alone* in creation and redemption to another movement of thought, *from* God *to* the world. Here the dynamic conception of God and the utter contingency of the world upon His Will, so powerfully advocated by the Reformers, released nature for empirical investigation out of itself while reverting to the Patristic insight that the rationality inherent in nature is conferred upon it by God's creation of the world out of nothing. This was the developing perspective in which the container notion of space that came out of the Renaissance and was taken up in different ways by Gassendi and Galileo came to be fully elaborated and built into the fabric of classical physics by Newton in a way in which, as Einstein himself has admitted, was the only possible and fruitful one at the time. Thus, over against Descartes' view of space as extension in which he modified the Aristotelian concept of the interdependence of container and contained matter through his idea of the coordinate system, and over against the thoroughly relational view of space advocated by Leibniz and Huygens in which they stood closer to certain aspects of Platonic and Patristic thought, Newton separated space from what happened in it, and put forward the idea of an infinite receptacle, formed by time as well as space, which he held to be the container of all creaturely being. He thus developed a dualism of space and matter, or volume and mass, in which space and time were given an absolute status independent of material existence but causally conditioning its character and qualities as an inertial system, thus making nature determinate and our knowledge of nature possible.

This calls for several comments from the point of view of our theological concern.

(i) Newton himself spoke of space and time as an infinite receptacle in terms of the infinity and eternity of God, for it is in God as in a container that we live and move and have our being.[1] Thus infinite volume is related in his thought to the Spirit of God and infinite time is identified with eternity—infinite space and time are in fact attributes of Deity. This is the way in which Newton operated within the movement of thought from God to the world and interpreted the idea that God confers rationality upon nature; but whereas Patristic and Reformed thought declined to give the receptacle idea a place in this, he held quite literally that God contains and comprehends the universe in Himself. But this joining together of God and the world by giving them space and time in common led finally into their separation.[2]

[1] I wish to refer here to the thesis of one of my students, Barry H. Downing, on 'Eschatological Implications of the Understanding of Time and Space in the Thought of Isaac Newton', 1966, for a full discussion of the way in which Newton related his physics to his theology.

[2] A similar point is made by C. F. von Weizsäcker about Nicolas of Cusa, *The World View of Physics*, p. 155.

(ii) In his doctrine of absolute space which is always similar and unmovable without relation to anything external, Newton returned to the Aristotelian and Medieval notion of a system of reference from a point of absolute rest. So far as the world was concerned this centre of immobility was the centre of gravity but since it was inevitably related to the immutability of God, Newton's physics and theology combined to arrest the great transition from Medieval to Modern thought, that is, from the ontic-static to the dynamic-noetic. It is not surprising, therefore, that within this cultural climate Protestant theology lapsed into a new scholasticism, freezing, as it were, the dynamism of the Reformers.

(iii) On the other hand, the fact that time was associated by Newton with space in the fundamental structure of knowledge meant that spatial extension in time and therefore history entered fully into the arena of scientific study. The effect of this, however, was paradoxical. Because time was inseparably bound up with space in conditioning matters of fact, the linear character of history, and thus the proleptic, predictive nature of events, had to be taken seriously. But because time and space were given an absolute status in a homogeneous and undifferentiated system, a static and necessary quality was imported into the notion of history. Here we have seeds of that fatal division between two kinds of history with which modern theology has recently been cursed.

(iv) Theologically the most important problem posed by Newton's thought was his association of space and time as an infinite receptacle with Deity, for that had the effect of reinforcing the dualism between space and matter which we have already noted. If God Himself is the infinite Container of all things He can no more become incarnate than a box can become one of the several objects that it contains. Thus Newton found himself in sharp conflict with Nicene theology and its famous *homoousion*, and even set himself to defend Arius against Athanasius. Here, then, we have a revival of the old Hellenic dualism that had long been nourished in the bosom of Augustinianism, that had taken a new form in Luther's distinction between the two kingdoms, and still another form in the dualism between subject and object in Cartesian philosophy; but now it was built into the structure of Western thought under the sanction of absolute space and time. Thus against all his religious intention Newton paved the way for the rationalist deism that developed in English thought and which had such a potent influence in Germany, reinforcing its own latent dualism. It should not be difficult to see that it is to this cross-fertilization of Lutheran by Newtonian thought that we owe some of the most persistent and deeply rooted problems that have come to the surface in Modern theology—but what made this cross-fertilization possible was the fact that Lutheran and Newtonian thought both operated with the old receptacle notion of space. We must restrict our discussion only to a few of the most far-reaching difficulties.

(a) We begin with the problems raised by dualism and the inertial relation it posited between absolute space and time and the creaturely world, which led so quickly into deism. The detachment of God and the questioning of any real interaction between Him and nature set the field for debate between believers and rationalists, theists and agnostics, but since it shifted the movement of thought back into the old perspective, from the world to God, it gave rise to a new natural theology, e.g. with Lord Herbert of Cherbury. That is to say, it sought to detach once again the rational structure of theology from actual knowledge and to work it out *in abstracto*, rather in the way in which Euclidean Geometry has been treated, as a detached science antecedent to the actual knowledge of the world gained through physics. The effect of this was to erect a rigid framework within which to receive and comprehend knowledge of God, but what it did was to develop a habit of projecting upon 'God' an objectifying form of thought rooted in our knowledge of the natural world, and thus to create objectivist concepts in theology corresponding to the static space of classical physics. These are the concepts that are so problematic for us today.

(b) The Lutheran principle *finitum capax infiniti* in this context of thought produced rather different fruit from that which was expected of it in the first place. Originally it had been applied to the humanity of Christ but His receptive capacity (*capacitas*) was held to be extended as His human nature was interpenetrated by divine majesty and power. This forced the Calvinists to ask what kind of humanity it is to which divine attributes can be ascribed, for even on Luther's own admission the humanity of Christ must be regarded as 'deified'. But if we remember that the humanity which the Son of God assumed is our humanity, that in which we share, who can stop short at applying divine attributes to the humanity of Christ and not apply them to humanity in general? This is what did happen, for the *capacitas naturae* defined in the humanity of Christ was extended to human nature as such, so that the *finitum capax infiniti* of Christ Himself was regarded only as a special and exemplary instance of man's own capacity for the divine. Thus it is difficult to dissociate the deification of man which we find in the eighteenth- and nineteenth-century German theology and philosophy from the Lutheran doctrine of *finitum capax infiniti*, although it took a rationalistic form in the capacity of the human spirit for the eternal and necessary truths of the reason.

(c) As the profound dualism this involved came more and more to the surface it undermined the reality of history and rendered the bond between Christianity and history ambiguous.

The very thing that Luther feared and fought against so hard, the transmutation of *est* into *significat*, of ontological relation into symbolic relation, came about. Although many influences converged at this point to produce it, the decisive disjunction came out of Lutheranism itself in the historic thesis of Lessing that 'the accidental truths of history can never become the proof of necessary truths of reason'. This did not mean for Lessing that the eternal truths of the reason are self-evident without regard to space and time, for it is only within space and time that we may apprehend them, but it did mean that since historical truth cannot be demonstrated, it cannot be used to demonstrate anything. There is, as he said, an ugly big ditch between them so that we must learn to leap from the accidental truths of history into quite a different kind of truth, discarding the historical as we go for it is but the external symbolic wrapping that is only of illustrative use. The effect of this, as Troeltsch affirmed, is to make the historical element in Christianity no more than the means of introducing the Christian idea into history, but once it is introduced it maintains itself on its own intrinsic resources. To investigate history, according to Lessing, is to appropriate it for ourselves and make it a part of our own experience as the particular symbolic point where the general truth of reason necessarily imposes itself upon us, and in the last analysis quite apart from any foundation in history. The effect of this was to reduce the historical Jesus Christ to a vanishing point, but Lessing's appeal to Luther indicates that, in his own thought at any rate, this had an inner connection with Luther's dynamic reduction of the presence of God to the mathematical point of 'for you'. Thus there grew up in modern Protestant theology a sharp antithesis between phenomenal events and eternal ideas, so that it came to be widely held that the spatial and temporal ingredients in theological concepts must be entirely discarded if we are to succeed in jumping over the chasm between them. But to retain spatial and temporal ingredients in the structures of our thought, as D. F. Strauss taught, is to remain stuck in *mythology*. In this view Strauss has had a massive following down to our own day, but it is high time that it was made clear that this concept of mythology rests upon the axiomatic assumption of a radical dichotomy between phenomenon and idea which is as impossible for Christianity as it is for modern science.

(d) It is in the connection of dualism with the principle of *finitum capax infiniti* that we may discern the theological roots of the so-called 'Copernican revolution' of Immanuel Kant. In that revolution the classical distinction between the *mundus intelligibilis* and the *mundus sensibilis* which appeared in his dissertation of 1770 was transformed into the dualism between 'things-in-themselves' and 'things-for-us', for he had come to the conclusion that the only knowledge we can reach is that of phenomena or of things in the mode in which they appear *to us* in our experience. We know only as we shape what we know for ourselves through the built-in capacities or *a priori* forms of the human understanding, so that we must regard things as they are in themselves as remaining quite unknowable. In other words, what Kant did was to apply to the human mind itself the receptacle notion of space which conditions and makes possible our knowledge of nature. But this transfer of limiting comprehension from God to man—for that is what it amounts to—implies that God in Himself remains quite unknowable beyond our intuitive representations, for He is not to be thought of as becoming the object of man's active reason, shaped by its power and held under its control like the phenomenal realities of our this-worldly experience. If God is 'known' it must be through some other way, by some kind of 'faith', but here Kant's radical dualism drove a deep wedge between faith and rational structure. It was partly in view of this that Schleiermacher, accepting the 'unknowability' of God, sought to find a way of integrating rational structure and the religious consciousness of the Christian community. But there is another aspect of Kant's thought which we must note. If space and time are *a priori* forms of man's sensory perception, then the point of absolute rest is transferred to a centre in man himself. His one fixed point is his self-understanding. Theologically this must mean that there can be no God for man outside of himself or independent of his consciousness, no divine Constant invariant with human determinations and valuations. Then the only God man can or will have is that which he postulates in his need and morally appropriates for himself. Lutheran theology then took its cue from Kant's practical reason in order to develop its old soteriological approach to Christology, and so Ritschl, rejecting metaphysical structures, sought to put Protestant theology on a new basis by interpreting Christ in terms of what He means for us in our moral need and understanding. Christ is here understood not as a 'thing-in-itself' but only as 'a thing-for-us'. However, this only accentuated the schizoid disjunction between history and ideas that had grown up and proved to be essentially ambiguous. Such a phenomenological abstraction of Jesus Christ from His living and actual situation in space and time, and therefore the observation, analysis and interpretation of Christian forms and structures isolated from their context in real history, could only lead to a reconstruction of Christianity either in accordance with a false ontology built out of mechanistic concepts or in a pietistic existentialism in accordance with man's religious self-expression.

(e) In the nineteenth century a significant change began to take place in the kind of questions being asked, which gave rise to a corresponding change in the notion of space. Hitherto scientific questions were concerned largely with concrete particularity, and yielded a view of matters of fact as discontinuous particles mechanistically related. The objectivist structures that came out of this were consistent with the notion of static space. Then questions began to be asked as to the fields of connection in which things are found, which yielded the view that they are not related like discrete bodies (although this is the way in which we observe them) but are connected up in a continuous flow of motion spreading through the universe. This gave rise to a relational and dynamic concept of space in which the old receptacle idea was finally shattered. That carries us into the space–time of relativity theory, but well before then in the nineteenth century absolute space had become highly questionable and static structures of thought began to be dismantled. What happened to Protestant theology in this context?

Undoubtedly the traditional theological structures that had been conditioned by the kind of objectivity developed in Newtonian physics began to crumble, and by the early twentieth century even the beautifully structured religious consciousness of Schleiermacher's *Glaubenslehre* had melted away. Protestant dogmatics was in rapid dissolution. Corroded by deistic dualism until it was detached from its proper objectivity in the interaction between God and the world, it was left hanging in the air, for even empirical science could not yet offer it a profounder objectivity upon which it might rely in place of the old one. Space was becoming elusive and time was vanishing as an indeterminate instant because of the loss of a fixed or definite centre of objective reference, while structures deprived of their particularities became vague symbolic manifestations of the eternal. Romantic idealism and romantic naturalism in different ways rubbed out differences and blurred distinctions without putting in their place objective connections of an ontological and a dynamic character that were needed. In these circumstances two courses were attempted. One was taken by the so-called 'Mediating School' of Lutheran theology, which sought to overcome the destructive tendencies at work in the nineteenth century by rebuilding upon the classical foundations of Patristic and Reformation thought, but assimilating into it accounts of the historical Jesus and His religious self-consciousness through the concept of *kenosis*, or self-emptying of the Son of God. But since it interpreted this still within the frame of the receptacle idea it only reproduced the old problems of the sixteenth century in a new form, and in the thought of Thomasius and Gess we find the earlier kenotic solutions in different if more philosophical dress. The other course taken led back to the Kantian idea that space and time are *a priori* forms of intuition outside the range of our experience and independent of it, or to the Hegelian idea that they are categories of the Absolute Spirit. This was not only to find its central focus in the human self-understanding but to develop a radical interpretation of Christianity cut off from the actual space and time of this world, in the mistaken view that one could by-pass the problems that worried the Kenoticists by holding the idea that the span of Christ's earthly life is only a mathematical point in relation to God although it is broken up into a sequence of time through the dividing and compounding operations of our human understanding. This was not so difficult to hold within the Newtonian perspective, for if space and time are the same and undifferentiated in all places and directions there are no boundaries in them. Thus by reference to absolute isotropic space and time it is possible to bring together in thought in the same instant what is observed or experienced as extrinsically differentiated, and in this way to link together the past and the present. This was already apparent in Newton's own typological interpretation of the Scripture.

Modern Protestant theology was apparently getting into a

neurotic condition, for behind its developing tendencies there appeared a schizoid anxiety to secure faith from the criticism of positivistic philosophy and science, to which it would be open if it were conditioned by space and time. But this could only bring Christian theology into contempt with Marxism and empirical science. Marxists reject both the view that space and time are empty receptacles independent of material processes and the view that they are only abstractions existing in the human consciousness, and hold that they have some form of objective reality through a profound relation with matter in motion. Hence they are unable to give any hearing to those who detach their message from space and time or preach only a Gospel of timeless events in which there is merely a tangential or paradoxical relation between history and the Kingdom of God. A Christ who is no more than a symbol for that kind of encounter is quite unintelligible. On the other hand, when modern theology ties itself to a fixed point in man's self-understanding so that man uses in his thought of God a system of reference which he determines or merely postulates for himself, this can only appear unintelligible and indeed quite arbitrary to modern science which must operate with constants irrespective of the observer or the choice of any particular system of reference.

Certainly, as Karl Heim has shown so clearly, modern science in the nineteenth century, when faced with similar difficulties, had to postulate an ideal point from which to penetrate behind the different coordinate systems of space in order to reach a higher level of abstraction within which to construct a unitary view, but it did so only in the search for a fuller and deeper *objectivity* and did not confound the 'as if' (*als ob*) operation in postulating this point with making it a substitute for objectivity. For natural science nature is and must be represented under the same laws for all observers, and, because of that *objective* understanding and procedure, marches forward as a unity, but modern theology that takes its cue from Kant in finding its centre of absolute rest in the mind of man and his self-understanding can only break up into as many theologies as there are theologians—yet this is not to engage in proper or scientific theology but to substitute in its place an ideological construct out of human subjectivity, individual or social. This is how the phenomenological approach to theology always ends up.

(f) Now we come to the problem of Bultmann's demythologizing programme, the most radical attempt in our day to think away space and time from the basic concepts of the Christian faith. Without doubt a real difficulty is being recognized and faced here. Too often the God of Newtonian Protestantism has been shaped according to the static, isotropic receptacle of space and time. Theological concepts of this kind are objectivist, rigid and closed. However, as we saw in the case of classical Lutheranism, it is no answer to the problem of spatial structures to make them timeless, yet it is into this old mistake that Bultmann has led so many of his contemporaries. At this point Bultmann has set himself in sharp opposition to Luther, for the *pro me* of Luther, what Christ has done 'for me' included and rested on an objective *pro me*, what Christ had done apart from me and outside of me; but Bultmann insists that the objective reference must be dropped altogether in order to get the meaning out of it 'for me'. This is the most ruthless radicalization of the *hoc significat* in place of Luther's *hoc est*, in the existentialist 'leap of faith' in which, like Lessing, Bultmann finally discards the place of historical facticity in the ground of faith. This is of course entirely consistent with Bultmann's deistic view of the relation between God and the world, which means that we can 'speak' about God only in terms completely detached from creaturely and this-worldly content or treat language about God as the paradoxical obverse about our self-understanding in this world. What is more, Bultmann is quite prepared with an obstinate courage to accept the consequences of regarding the present as a timeless instant, viz. that the past has vanished for ever and as such can have no meaning for us, which cuts away the historical Jesus, and that the future offers us no existence, which cuts out the hope of the resurrection. All that matters is what is 'for me' here and now.

In view of the preceding discussion only two things need to be added. First, just as the linguistic philosophy so dominant in England now appears as a strange anachronism, since it is evidently rooted in the outmoded particulate view of nature, so Bultmann's flight from space and time can only appear as an obsolescent survival out of the uncertainties and anxieties of the late nineteenth century. Of course if he still operates with the Lutheran receptacle notion of space—and, like Tillich, Bultmann has rejected the '*extra Calvinisticum*'—then he is already mythologizing the Incarnation by assuming it to mean that the Son of God is received and contained within the dimensions of a human body. However, what makes any idea of God's presence and activity within cosmic space and time mythological for Bultmann, is his deistic assumption that God does not interact with this world, which he regards as a closed continuum of cause and effect, not to speak of the antithesis between a noumenal world of 'things-in-themselves' and a phenomenal world of 'things-for-us' which he inherits in different ways from Kant, Schleiermacher and Ritschl. In the second place, it may be pointed out that his critique of the whole structure of Nicene theology rests upon a strange confusion of questions and languages. This may be illustrated from a similar problem in physics noted by Jean Charon.[1] 'Relativity

[1] Jean Charon, *Man in Search of Himself*, p. 116. See also pp. 58 f., and 117 f.

theory', he says, 'represents the universe in terms of a symbolic language, and enables us to visualise this universe as a whole, in space and time, as the surface of a sphere.' What then are we to make of questions as to what is outside this sphere? They are simply meaningless. Relativity theory defines the whole universe, and it is nonsensical to ask what is beyond everything. Moreover, the questions are framed in a very different language, relating to what we observe through the senses, and cannot rationally apply to relativity theory framed in language that carries us quite beyond what is observable. They are therefore impossible. It seems to me that the questions Bultmann has raised about theological structures are nonsensical precisely in this way, for he is bringing two different languages together at the same time, which cannot be done without contradiction and absurdity. Even when these languages intersect it is still nonsensical to mix or confuse them.

How then are we to assess the problem of spatial concepts in Reformation and modern theology? Undoubtedly Luther's insight was right in trying to relate the ontological and the dynamic ways of thinking of the real presence and the incarnation. Luther took an enormous stride in the move from Medieval to modern times which we have spoken of as from the ontic–static to the dynamic–noetic, but history has also justified his fears that they might fall apart—yet it was Lutheran unwillingness to give up the receptacle view of space that helped to create the problematic state of affairs in which the twin approaches fell disastrously apart. Had Lutherans followed Patristic and Reformed theology in rejecting the receptacle idea and developed a relational and differential concept of space, instead of extrapolating Medieval problems into the modern world, the story of Protestant theology might have been very different. But, all the way through, it was German theology that created the pace, with the result that its problems were everywhere injected into Protestant thought. Certainly the task of Modern theology clearly lies in a unitary movement of thought in which the ontological and dynamic approaches are brought together. If so we must set ourselves to the task of developing appropriate logical tools and an apposite theological language for this purpose—a task not altogether unlike that on which contemporary physicists and biologists are engaged. This will require us to work in close association with them, however, in clarifying the different metaphysical and logical levels of thought within each science and exploring how they are and must be coordinated within the same universe of knowledge.

CHAPTER 3

INCARNATION AND SPACE AND TIME

By the Incarnation Christian theology means that at a definite point in space and time the Son of God became man, born at Bethlehem of Mary, a virgin espoused to a man called Joseph, a Jew of the tribe and lineage of David, and towards the end of the reign of Herod the Great in Judaea. Given the name of Jesus, He fulfilled His mission from the Father, living out the span of earthly life allotted to Him until He was crucified under Pontius Pilate, but when after three days He rose again from the dead the eyes of Jesus' disciples were opened to what it all meant: they knew Him to be God's Son, declared with power and installed in Messianic Office, and so they went out to proclaim Him to all nations as the Lord and Saviour of the world. Thus it is the faith and understanding of the Christian Church that in Jesus Christ God Himself in His own Being has come into our world and is actively present as personal Agent within our physical and historical existence. As both God of God and Man of man Jesus Christ is the actual Mediator between God and man and man and God in all things, even in regard to space–time relations. He constitutes in Himself the rational and personal Medium in whom God meets man in his creaturely reality and brings man without, having to leave his creaturely reality, into communion with Himself.

This is an utterly staggering doctrine. It does not mean, of course, that God has resolved Himself wholly into what He was not or that He has merged His eternal reality entirely with the creaturely reality of man. Nevertheless it is to be taken in all its serious intention to mean that the Son of God has become man without ceasing to be the God He ever was, and that after the Incarnation He is at work within space and time in a way that He never was before. As soon as we talk like this, however, as Origen was quick to point out in the *De Principiis*, or even say about the Son that 'there never was a time when He did not exist', we are using terms 'always', 'has been', 'when', 'never', etc., which have a temporal significance, whereas statements about God, Father, Son and Holy Spirit, must be understood to refer to what transcends all time and all ages, and all eternity, since even our concept of eternity contains a temporal ingredient. How then can we speak about the Incarnation as act of God in this way without illegitimately projecting our creaturely time into God?

This question must be pressed even further, for it is surely relevant to everything we say about God. All our speech about Him is creaturely and cannot be cut loose from its creaturely content without ceasing to be human speech altogether, but it is not for that reason false as if it could have no authentic relation to the reality and intelligibility of God. Here we are up against one of those ultimate boundaries in thought such as we reach when we ask a question as to the rationality of the universe: not only do we have to assume that rationality in order to answer the question but we have to assume it in order to ask the question in the first place. We cannot meaningfully ask a question that calls in question that which it needs in order to be the question that is being asked. We cannot step outside the relationship to the rationality of the universe in which we find ourselves without stepping outside of rationality altogether. Before the question as to the relation between our knowing and ultimate rationality we cannot but stand in awe and acknowledgement, and can ask our questions rightly only within the actuality of that relationship. That is the problem that arises in our knowledge of God, in determining the relation between our thought and speech and the reality of God about whom we think and speak, for in the nature of the case we operate with an incongruence between our knowing and the divine Object of our knowing. We can only stand before Him in wondering awe, acknowledging that in our knowing of Him we are unable to reduce the relations between our thought or speech and the reality of God to relations of thought or speech, and so we have to be on our guard against asking the meaningless kind of question that presupposes we can.

That is why, with reference to our human inability to overcome the incongruence between human knowing and the reality of God, Karl Barth has asked: 'How do we come to think, by means of our thinking, that which we cannot think at all by this means? How do we come to say, by means of our language, that which we cannot say at all by this means?' If there is to be real knowledge of God there cannot but be an incongruence between God as the known and man as the knower, but if that knowledge is to take place it must rest upon the reality and grace of the Object known—just as in all true knowledge, where we are unable to reduce the relation between our thought and speech and the reality of things to relations of thought and speech, we nevertheless allow their reality to shine through to us and act upon us. And so Barth rightly insists that in the knowledge of God we cannot raise questions as to its reality from some position outside of it.

We are, therefore, restricted to the sharp alternatives: *either* to be entirely silent, i.e. not even to venture the sceptical question, for at this point, as in regard to the rationality of nature or the laws of thought, it could only be a contradictory and nonsensical movement of thought; *or* to ask questions only within the circle of the knowing relationship in order to test the nature and possibility of the rational structures within it. If we take the first alternative we must also reject every form of science since all scientific questions presuppose the rationality of that into which they inquire. But if we take the second alternative we must insist upon rigorous scientific procedure in which we never ask questions outside the knowing circle, for that would mean asking irrelevant questions that were not matched with the nature of the subject-matter. Here we seek to bring to light the rational grounds upon which our knowledge claims to rest, either to establish it evidentially upon those grounds in such a way as to exhibit a thoroughgoing consistency between our understanding and that into which we inquire, or to use the rationality that comes to light and the coherence of our operational structures to enable us to discriminate between realities and fictions.

Now when within this approach we inquire into the Incarnation and take up again Origen's question as to the legitimacy of using terms with spatial and temporal ingredients to speak of God who in His own essence does not exist in a spatial or temporal relation to the creaturely world, we find that any meaningful answer must bring together the doctrines of the Incarnation and Creation. If our human and creaturely language about God is to have any real intention terminating objectively upon Him, it can only be on the ground of His interaction with the world He has created and within the relation that He has established between it and Himself. Put the other way round, this means that statements about acts of God in the Incarnation imply and demand statements about the creation of the universe and the unique relation in which as Self-existent Being the Creator stands to the creation, for it is within that relation that we as God's creatures belong and within it alone that we can conceive and speak of God in any true way. It also means, however, that our ordinary terms and concepts formed in the first place for purposes of intra-mundane knowledge must be stretched and adapted in accordance with the relation between the creation and the Creator. If they are to fulfil their intention in speaking about God they must be made to point beyond the spatial and temporal limits of the contingent world in a relation of transcendental reference, yet not in such a way that they are detached from the contingent world out of which that reference is made. If they were cut off like that they could not mean anything to us at all, for they would not be grounded with us in the same relation which, as human creatures within this world, we have with the Creator of all things.

Since our concern here is not with Incarnation and Creation as complete doctrines, but with Incarnation in its relation to Creation in respect of the questions of space and time, let us recall the principal conceptions of *space* and *time* that have arisen in the history of thought which theology has had to use and adapt for its own specific purpose, remembering also that it is largely to Christianity that we owe the important place given to *time* in the development of Western thought. These are the *receptacle notion* of space, or of space and time, and the *relational notion*. The receptacle notion which has certainly predominated in ancient and modern times crops up in two different forms, the finite receptacle and the infinite receptacle. In the Greek world both had to do primarily with space. When space was thought of as a finite receptacle it was not regarded as independent of that which went on in it, for anything other than determinate existence was held to be unthinkable and unintelligible. In Aristotelian and Stoic thought this inevitably carried with it the idea of a finite universe and a finite god. This receptacle concept of space was rejected by Patristic theology, but when it entered into Western thought with Aristotelian physics and philosophy under the category of magnitude, apart from the conception of time, it tended to encourage a *deus sive natura* view of the relation between God and the world but at the same time led to their separation in a sharp distinction between nature and supernature and consequently to a corresponding distinction between natural theology and revealed theology. When, on the other hand, space was thought of as an infinite receptacle, which we find in the teaching of Atomists and Pythagoreans alike, it was regarded as independent of what went on in it, and was often spoken of as 'the great void'. This tended either towards materialism and atheism or to a conception of a detached unknowable deity calling for mystical, non-rational communion. This receptacle notion of space was also rejected by the Church Fathers, but it entered Renaissance thought through the Florentine Academy, and is found in the thinking of people as different as Patritius, Gassendi and Galileo. From them it was taken up, modified, and elaborated, by Newton who gave it, together with time, an absolute status independent of material bodies but causally conditioning their character as an inertial system, thus making nature determinate and our knowledge of nature possible. It was thus that there was built into the heart of Western science and philosophy a far-reaching dualism between space and matter, or volume and mass, which was to have its effect in the fatal division between two kinds of history. Moreover, since space and time were regarded by Newton as attributes of God, constituting the infinite Container of all creaturely existence, this view also tended towards a *deus sive natura* conception of the relation between God and the world by giving them space and time in common. But as a matter of fact it led to a separation of God and the world for it gave rise to a deistic dualism, and so to a new form of natural theology more rationalistic than its Medieval counterpart.

The other principal conception of space and time is the relational idea which was given its supreme expression in the space–time of relativity theory when Einstein, following out a line of thought from four-dimensional geometry, found he had to reject the notion of absolute space and time both as taught by Kant, for whom they were *a priori* forms of intuition outside the range of experience, and as taught by Newton, for whom they formed an inertial system independent of material events contained in them but acting on them and conditioning our knowledge of the universe. This has had the effect of shattering the receptacle idea and of undermining the radical dualism to which it had given rise in modern philosophy and theology as well as science. The relational view had been anticipated by Leibniz and Huygens in certain respects, but its early roots may be traced in Plato's *Timaeus*, whose account of space is as much epistemological as cosmological and is equally applicable to time, and in the teaching of Theophrastus, the pupil and critic of Aristotle, who regarded space as a system of orderly interrelations between the positions of bodies. These ideas had their measure of influence on the Greek Fathers, together with the Stoic principle that space must be understood in terms of an active principle that makes room for itself in the finite universe, but they refused to admit the receptacle notion of space into theology on the ground of the biblical teaching about God as the transcendent Source of all being whom the heaven of heavens cannot contain. In the light of God's creation of the world out of nothing, His interaction with nature, and the Incarnation of His Creator Word, they developed a thoroughly relational conception of space and time in which spatial, temporal and conceptual relations were inseparable, for like Plato they held that the basic problem was more an epistemological than a cosmological one. This Patristic understanding of space and time was appropriated at the Reformation by Reformed and Anglican theology, but so far as the development of Protestant theology is concerned it was smothered by the dominance of Newtonian and Lutheran thought. Now that the receptacle notion of space and time has broken down, although the confusion to which it gave rise, not least in the understanding of history, is still rife, we need to rethink the essential basis of Christian theology in the relation of the Incarnation to space–time, and to think completely away the damaging effects of a deistic relation between God and the universe.

The Christian doctrine of Creation asserts that God in His transcendent freedom made the universe out of nothing, and that in giving it a reality distinct from His own but dependent on it He endowed the universe with an immanent rationality making it determinate and knowable. Over against the creation God remains quite free in His eternal Self-existence and cannot therefore be known in the determinate way in which created things are known, but the creation also remains free in its utterly contingent character and is therefore to be known in its natural processes only out of itself. When considered theologically, that is from a centre in God, the world is to be understood as subsisting in His creative Word, for God continues to maintain it in being through His freedom to be present in it and to realize its relation as created reality towards Himself the Creator. When considered naturally, that is from a centre in nature itself, it is to be known and understood in so far as we penetrate into its inherent intelligibility and succeed in giving it intelligible representation. It is owing to this relation of creative freedom between God and the creation, and of contingent freedom between the creation and God, that nature by itself speaks only ambiguously of God, for while it may be interpreted as pointing intelligibly beyond itself to God, it does not permit of any necessary inferences from its contingence to God. Thus the fact that the immanent rationality of the universe is unable to give any final account of itself is the obverse of the fact that the rational connection between the creation and God is grounded in God alone, and does not rest partly in God and partly in the creation. It is for this very reason, namely that the creation acquires its rationality in God's creative comprehension of it, that it is constituted and enabled to be the rational medium through which God speaks to us and makes Himself known, and in which once and for all His own eternal Word has become man. At the same time this lets us understand why the conceptual and linguistic media employed in the knowledge of God through Jesus Christ cannot be detached from our knowledge of this world in its creaturely rationality, and why whenever through our scientific questioning that creaturely rationality comes to view we cannot help but discern the transcendent rationality of God shining through it.

It is in this complex of interrelations between the Incarnation and creation that we have to consider the relation of the Incarnation to space and time. The creation of the universe out of nothing implies the absolute priority of God over all space and time, for space and time were produced along with the creaturely world as orderly functions of contingent events within it. They cannot therefore be bracketed with God after the manner of Newton, for the doctrine of God as the Maker of all things, visible and invisible, excludes any mythological synthesis between God and the universe, God and nature, or between the divine reality and this-worldly reality. God stands in a transcendent and creative, not a spatial or temporal, relation to the creaturely world. Hence even the relation between the actuality of the Incarnate Son within this world of space and time and the Father from whom He came cannot be spatialized or temporalized. Nevertheless, since the whole realm of space and time is maintained by God as the object of His creative knowledge and power, space and time are to be conceived as a continuum of relations given in and with created existence and as the bearers of its immanent order. Apart from space and time nature would be indeterminable and unintelligible, for it would have no sequences or patterns of change and no series of continuous coherent structures and would thus be incapable of any kind of meaningful formalization. It is to space and time, therefore, that we have to look for the determinate and intelligible medium within which God makes Himself present and known to us and within which our knowledge of Him may be formed and grounded objectively in God's own transcendent rationality. This is why any so-called 'demythologizing' that aims at stripping away the spatial and temporal ingredients from our theological concepts and terms, or reinterprets the spatio-temporal structure of the Creed in such a way as to have the same effect, in a spaceless and timeless relation to God, can only lapse into irrationality and meaninglessness.

How then are we to conceive the relation between the transcendent rationality of God independent of space and time and the rationality immanent within the space–time structures of this world? This question becomes very acute when we think of the Incarnation as the condescension of God to come Himself into the determinations, conditions and conceptualities of our world through the assumption in Jesus Christ of particular physical existence in space and time. Do we give this responsible consideration if we interpret it merely along the line of the principle enunciated by Clement of Alexandria that God is far off in respect of His essence or being but very near in respect of His power? While this has a very important point to make, can we be satisfied with construing the presence and acts of God among us only in dynamic and not at all in ontological terms, or with the radical dualism between the intelligible and sensible realms which is implied in Clement's thought? This has undoubtedly been the tendency in modern theology when it holds that God is related to our world only tangentially at mathematical points, so that His presence and acts within the cosmos can be spoken of only paradoxically, but that means that man's knowledge of God cannot be grounded cognitively and objectively in God for it only touches the world of divine reality asymptotically. This could only be the end of the road for modern theology, for such a retreat into spacelessness and timelessness is a retreat into irrationality. Whatever we do, we cannot contract out of space and time.

If then we accept the need to take space and time seriously there appear to be four ways in which we can think of the Incarnation.

(i) The coming of the Son of God into space and time may be regarded as His entry into a finite receptacle. Theologians who have held this view have usually operated with some form of the Aristotelian notion of place as a limited and limiting container—this is what we find among Thomists and Lutherans. But whereas St. Thomas brought into play at this point the Patristic teaching that the Son of God became man without leaving the throne of the universe and so modified the receptacle idea by taking the lid off it, as it were, Luther declined to do that, with the intention of rejecting monophysitism, but he had to bring in a strong form of the *communicatio idiomatum* in order to get out of his difficulties. This only meant that he had to extend the human receptacle, e.g. in the ubiquity of the body of Christ, in order to make it contain His omnipresence, and so monophysitism came in by the back door. *Finitum capax infiniti.* After Luther the struggle to retain the humanity of Christ as the earthen vessel into which the Son of God emptied

Himself gave rise to kenotic theories which led either to a restriction of the Deity of Christ or to the imposition on the Son of God of an objectifying form of thought. The latter became even more pronounced when with the Kantians the receptacle idea was transferred to the human mind. Thus in one way or another this view destroys itself or collapses through its inherent contradictions.

(ii) The second view operates with the theory that God Himself is the infinite receptacle, *infinitum capax finiti*. This was the idea advocated by Newton and integrated with the structure of classical physics and its dualism between absolute space and time and material existence. On these terms it proved impossible to conceive of the Incarnation in any serious way, for if God is the infinite container of all being He can no more be one of the particular beings He contains than a vessel can at the same time be one of the things contained in it. Hence this view led right away into Arianism while the dualism that lay at its heart rapidly developed into sheer deism in its classical form. Historically and theologically the significance of these ideas was their cross-fertilization with German Lutheran thought giving rise to the profound split between God and the world, noumenon and phenomenon, *Geschichte* and *Historie*. The continuation of this dualism in modern theology is a crude anachronism, for it belongs to a cosmological theory that is no longer possible, but it was a theological impropriety in any case, i.e. a pseudo-theological construct deriving from an alien concept.

It is evident, then, that both forms of the receptacle idea of space and time must be rejected, although it will take a long time to clear up all the damage they have caused.

(iii) Now we come to the view advocated by Origen in the ancient world and which has an interesting counterpart in the modern world, e.g. in the thought of E. A. Milne. Here as in

Patristic theology generally space and time, and all the orderly relations within the universe, are understood through reference to the creative and all-embracing power and activity of God, but according to Origen God has endowed the creation with His own rationality. This meant, as Origen pointed out, that God has limited Himself in the limitation of creation through its subjection to His own self-comprehension. This had the advantage of regarding space and time as the rational medium through which God is revealed and through which, therefore, theological statements may be traced back objectively to their intelligible ground in the Being of God, but it also had the effect of delivering men from a view of the universe ultimately subjected to futility and purposelessness through being unable to explain its own rationality. In Origen's thought this implied the eternal coexistence of the universe in the mind of God, although he sought to guard himself against difficulty by asserting the priority of God even over 'eternal intelligence'; but in the thought of E. A. Milne this way of connecting the rationality of nature with the transcendence of God implies a necessary relation between them. This was in line with Milne's unitary theory of nature and the possibility of a purely mathematico-deductive account of the laws of nature which he shared with Eddington, but which does not appear to do justice to the contingence of nature. Theologically it provided for Milne rational argumentation for God, but it meant the limitation of God since He is not free from a relation of necessity between His transcendent rationality and that of the material universe. Moreover, this would seem to lead us into a difficult situation in which we would be unable in the last resort to distinguish divine objectivity from worldly objectivity.

(iv) The fourth view appears in different forms in Patristic theology, in the thought of Anselm, Duns Scotus, Pascal and

Karl Barth. This is the view that the structures of space and time are created forms of rationality to be distinguished from the eternal rationality of God. In creating and knowing them God remains free from any necessity in the relationship, although they remain grounded in the Supreme Truth of His Being. Hence the Incarnation of the Son of God in the realm of space and time means that He assumes created truth and rationality and makes them His own although He is distinct from them. As embodying in Himself the Supreme Truth God remains in a relation of freedom and transcendence and is not at our disposal in the conceptualism and necessity of rational argumentation. The difficulty of this view, at least in some of its forms, is that it opens up the way back to the idea that in Himself God remains finally inscrutable. This is what happened in the development of Occamism when human knowledge of God was made contingent upon His Will and not connected with the Truth of His Being or even primarily with His Mind. This could not have arisen in Anselm's thought for he held that while God is ineffably transcendent over all our conceiving of Him, yet our conceiving, when true, arises under the compulsion of the divine Being. Created rationalities thus embody an element of necessity, i.e. their impossibility of being other than they are, in relation to the Supreme Truth of God. It is in virtue of this ontic necessity that theological knowledge cannot be arbitrary, but must be controlled through objective relation to the Truth of the divine Being.

Now if we accept Anselm's account of this relationship there ought to be a closer connection between the created conceptualities of theology and those of natural science than we find in the thought of Karl Barth, even when we take into account the necessary shift in meaning that theological concepts involve in accordance with the nature of their divine Object. There is clearly a very difficult problem here in the relation of freedom and necessity between God and the rationality of the creaturely world. Perhaps we may get nearer to the question by a consideration of Leibniz's idea that the real world in which we live is the best of all possible worlds. Like Anselm he held that God's rationality is bound up with His being who He is and not another. That is the Truth or Necessity of His Self-existent Being. The world does not have that kind of necessity. Since it was created out of nothing, it might have been quite different from what it is, but now that it has come into being it has a contingent necessity in that it cannot not be what it now is. Considered in itself, then, there is only *the* world, this world that has come into being, but considered from the side of God's creation it is only one of all possible worlds. Thus we must think of God's relation to the world in terms of an infinite differential, but we must think of the world's relation to God in terms of a created necessity in which its contingence is not negated. This relationship between divine freedom and contingent necessity in the world, as von Weizsäcker has shown, is exactly analogous to that found in the variational principles of physics,[1] as is apparent when we apply Fermat's principle, that light takes the shortest path between two points, to the formulation of natural law, for the selection of one possibility as the real one thereby stamps the others as really impossible. Since natural laws arise in this *a posteriori* manner they cannot be converted into *a priori* principles enabling us to argue prescriptively beyond what is empirically and contingently given, that is beyond the created structures of which natural laws are *post hoc* features.[2]

The implications of this are highly significant for our understanding of the relation between God and the world in

[1] C. F. von Weizsäcker, *The World View of Physics*, pp. 182 ff., 186 ff.

[2] See D. M. Mackay, 'The Sovereignty of God in the Natural World', *Scottish Journal of Theology*, 21.1, p. 18.

Incarnation as well as creation. The conception of an infinite differential in the rationality of God allows us to say that God is free from any necessity, spatio-temporal, causal or logical, in His relationship with the creation, without making Him arbitrary and therefore inscrutable. On the other hand, since the concepts and laws that arise within space and time are in the first instance concepts and laws of created realities we may not apply them in an *a priori* and necessary manner to God. Yet since they actually derive from His creation we must not think of the Incarnation as an intrusion of the Son of God into the determinations and conditions of space and time, or of His miraculous acts within them as in any way an abrogation of the space–time structures of this world that we call natural laws. Rather is the Incarnation to be understood as the chosen path of God's rationality in which He interacts with the world and establishes such a relation between creaturely being and Himself that He will not allow it to slip away from Him into futility or nothingness, but upholds and confirms it as that which He has made and come to redeem. Thus while the Incarnation does not mean that God is limited by space and time, it asserts the reality of space and time for God in the actuality of His relations with us, and at the same time binds us to space and time in all our relations with Him. We can no more contract out of space and time than we can contract out of the creature–Creator relationship and God 'can' no more contract out of space and time than He 'can' go back on the Incarnation of His Son or retreat from the love in which He made the world, with which He loves it, through which He redeems it, and by which He is pledged to uphold it—pledged, that is, by the very love that God Himself is and which He has once and for all embodied in our existence in the person and being of Jesus Christ. That is the infinite freedom and the unique kind of necessity that hold between God and the world, which not only preserve its contingence but which so ground it in the being and rationality of God as to provide for us in our creaturely existence an intelligible medium and an objective basis for all our relations with God. Therefore, now that the Incarnation has taken place we must think of it as the decisive action of God in Christ which invalidates all other possibilities and makes all other conceivable roads within space and time to God actually unthinkable. In this way the Incarnation together with the creation forms the great axis in God's relation with the world of space and time, apart from which our understanding of God and the world can only lose meaning.

Now when we come to formulate our theological understanding of space–time in relation to the Incarnation we are faced with a problem posed for us by the way in which many people today want to conceive of natural law, especially those who operate with a purely instrumentalist view of science. For them natural law is defined in terms of what we can do with things, which is to invert what it really is, viz., a rational representation of interconnections and regularities in the events themselves. Thus it amounts to a pragmatic Kantianism in which technology is substituted for pure science. This would appear to be quite impossible in view of our modern understanding of space and time as properties of bodies in motion, even though we must formulate what we understand by this in operationally defined terms. In more technical language, the flow of time and the extent of material bodies depends on the velocity at which those bodies move, so that the geometrical structures of space–time change according to the accumulation of mass and the field of gravitation caused by it. Since the four-dimensional geometries have destroyed not only the view that space and time inhere absolutely and statically in the universe but also the view that they are mere abstractions existing in our human consciousness, space and time must be conceived as the structural functioning of contingent events inseparably and objectively rooted in them. This means, for example, that space must be defined in terms of bodies or agents conceived as active principles, making room or creating space for themselves in the universe, but must be understood in accordance with the nature and movement of the bodies or agents in question, and not simply in terms of our own operations. Of course there will be a kind of space to be defined in terms of our own nature and movement, in interaction with other agents and the world around us, the organized structure of space–time in which we live our lives, but all this is in line with the variational principles which we noted earlier.

The fact that four-dimensional geometries are not just other ideal possibilities inventively thought up (even though mathematicians happened on them in the first place), but involve a profound correlation between abstract conceptual systems and physical processes, has considerable epistemological implications for theological as well as natural science, if only because it yields the organic concept of space–time as a continuous, diversified but unitary field of dynamic structures, in which the theologian as well as the natural scientist is at work. Since this gets rid of the old dualisms between material existence and absolute space and time, or between nature and supernature, it is no longer possible to operate scientifically with a separation between natural theology and revealed theology any more than between geometry and physics. In physics, this means that geometry cannot be pursued as an axiomatic deductive science detached from actual knowledge of physical processes or be developed as an independent science antecedent to physics, but must be pursued in indissoluble unity with physics, as the science of its inner rational structure and as an essential part of empirical and theoretical interpretation of nature. In theology, this means that natural theology cannot be undertaken apart from actual knowledge of the living God as a prior conceptual system on its own, or be developed as an independent philosophical examination of rational forms phenomenologically abstracted from their material content, all antecedent to positive theology. Rather must it be undertaken in an integrated unity with positive theology in which it plays an indispensable part in our inquiry and understanding of God. In this fusion 'natural' theology will suffer a dimensional change and will be made *natural* to the proper subject-matter of theology. No longer extrinsic but intrinsic to actual knowledge of God, it will function as a sort of 'theological geometry' within it, in which we are concerned to articulate the inner material logic of knowledge of God as it is mediated within the organized field of space–time.

From this it follows that instead of the false dualist approach in which we would be forced to interpret the space–time of the Incarnation in concepts that we had already developed independently in some area of natural knowledge, we must seek to build up a specifically theological interpretation, with its own apposite forms of thought and speech, within the unitary interaction of God with our world in creation and Incarnation, and within the unity of the rational structures that result from that interaction. This will involve what I have just called 'theological geometry', but that would be a misleading expression if it were taken to mean that we are to extrapolate into theology the operational principles of physics, even though theology and physics share the same space and time of this world. And it cannot mean that, if space and time have to be understood in terms of the nature and movement of bodies and agencies. But it does mean that if we are to think in this way of the space–time of the Incarnation we must allow the Incarnation itself to create for us the field of organic connections within which we are to develop our thought and language about it. Jean Charon has recently told us, in a way that reminds us of Michael Polanyi, that biological science will not achieve the advance that physics has made if it remains stuck in mechanistic concepts and language, but must adopt the notion of a field with its own characteristic structure in order to begin to be in a position to make effective use of the physicists' space–time.[1] That warning is no less applicable to theological science—the field that we are concerned with is surely the interaction of God with history understood from the axis of Creation–Incarnation, and therefore without the fatal split between *Historie* and *Geschichte* that has resulted from Newtonian physics and deistic natural theology. Instead of that dualism with its confusion of different languages we must learn to ask questions in two opposite directions at the same time, developing a relational and differential understanding of space and time in accordance with the nature and acts of God and in accordance with the nature and acts of man. Our understanding of this field will be determined by the force or energy that constitutes it, the Holy and Creator Spirit of God. It will certainly involve connections that are impossible from a merely dualistic point of view, or impossible if transferred outside this specific field—the same impossibilities would arise, of course, if we confused different field-languages in any area of science, especially that designed for what is inherently non-observable and that which arises out of observational experience.

The rejection of radical dualism, however, must not be taken to imply the advocacy of any oneness or even any proportion between God and the world, but rather the rejection of a deistic disjunction between them. God's interaction with the world He has made maintains a proper dualism between them.

[1] Jean Charon, *Man in Search of Himself*, ch. 4, pp. 68 ff.

Hence there can be no resolving of divine transcendence into this-worldly transcendence or any merging of the divine reality and this-worldly reality on the same horizontal level, even if that be regarded as an inclined plane. In Jesus Christ the divine reality intersects this-worldly reality like an axis, so that if our language about God who cannot be observed and our language about the world which can be observed, must not be confused, it is because they intersect at decisive points, and not because they are merely the obverse of each other or because they are merely parallel to one another. The interaction of God with us in the space and time of this world sets up, as it were, a coordinate system between two horizontal dimensions, space and time, and one vertical dimension, relation to God through His Spirit. This constitutes the theological field of connections in and through Jesus Christ who cannot be thought of simply as fitting into the patterns of space and time formed by other agencies, but as organizing them round Himself and giving them transcendental references to God in and through Himself. He generates within these connections His own distinctive and continuous 'space–time track', and forms a moving and creative centre for the confluence of world-lines within the plenum of space–time. The movement of eternity into time in Jesus Christ has the effect of temporalizing space and spatializing time in an orderly continuum of successive patterns of change and coherent structures within which God may reflect and fulfil His own creative and redemptive intentionality. It is not a movement that passes over into these structures or gets stuck in them, for it continues to operate livingly and creatively in space–time, travelling through it, fulfilling the divine purpose within it and pressing that fulfilment to its consummation in the new creation. It is therefore a teleological as well as an eschatological movement, in which the incarnate Word calls space and time, as it were, into contrapuntal relation to the eternal rationality of God, which because of its infinite differentiality does not override but maintains and fulfils the freedom of the created order.

This gives us, in the language of the physicists, 'an organized structure of space–time', but one that is made and kept *open* for a transcendent rationality that preserves its creatureliness and gives it meaning. This does not import the slightest rejection of this-wordly realities or the reduction of history to vanishing points in timeless and spaceless events, but rather the affirming and confirming of creaturely and historical existence in all its spatio-temporal reality by binding it to an eternal reality beyond the meaninglessness and futility to which it would be reduced if it were abandoned by God to itself. Moreover since the field of organized space–time is to be referred not to a centre of absolute rest in an unmoved Mover, but to the dynamism and constancy of the living Creator, it is linked with an inexhaustible source of possibility, because of which created and historical existence is so full of endless spontaneity and surprise that there are *no rules* for the discovery of its secrets. In virtue of the relation of space–time structures to the infinite differentiality of God they are essentially open and not closed to the humble inquirer who accepts and waits upon their rationality, but because of that relation he cannot infer genuinely new knowledge of the world even from what he has already learned and cannot predict *a priori* what he will discover, and only when he does discover what is new can he connect it up with what he has already learned.[1] This applies above all to the knowledge of God in Jesus Christ for in Him we find that God always takes us by surprise and we can know the Father only by *following* the space–time track in truth and life that is Jesus Christ. Hence the principle formulated by St. Thomas Aquinas: *Christus qui, secundum quod homo, via est nobis*

[1] Cf. E. H. Hutten, *The Ideas of Physics*, p. 64.

tendendi in Deum. Since in Jesus Christ the eternal Son has entered within the contingence of the created order, making it His own, He may be known only in and through its creaturely freedom and spontaneity, and therefore not in any *a priori* manner; yet since His created actuality in this world results from the transcendent freedom of God in condescending to become man, we cannot know Him truly except in accordance with that divine movement in the Incarnation and on the ultimate ground of God's unconditional self-giving in Jesus Christ. It is out of His fullness that we receive, grace for grace.

The world, then, is made open to God through its intersection in the axis of Creation–Incarnation. Its space–time structures are so organized in relation to God that we who are set within them may think in and through them to their transcendent ground in God Himself. Jesus Christ constitutes the actual centre in space and time where that may be done. But what of the same relationship the other way round, in the *openness of God* for the world that He has made? Does the intersection of His reality with our this-worldly reality in Jesus Christ mean anything for God? We have noted already that it means that space and time are affirmed as real for God in the actuality of His relations with us, which binds us to space and time, so that neither we nor God can contract out of them. Does this not mean that God has so opened Himself to our world that our this-worldly experiences have import for Him in such a way, for example, that we must think of Him as taking our hurt and pain into Himself?[1] This is what we cannot do from the approach of deistic dualism—why, for example, Schleiermacher could not hold that God is merciful and why Bultmann cannot allow that the love of God is a

[1] Cf. K. Woollcombe, *Scottish Journal of Theology*, vol. 20.2, 1967, pp. 129 ff.

fact within the cosmos. Thus it would appear that the question as to the impassibility of God is the question as to the actuality of the intersection of God's reality with worldly reality, and as to the depth of its penetration into our creaturely being. If God is merely impassible He has not made room for Himself in our agonied existence, and if He is merely immutable He has neither place nor time for frail evanescent creatures in His unchanging existence. But the God who has revealed Himself in Jesus Christ as sharing our lot is the God who is really free to make Himself poor, that we through His poverty might be made rich, the God invariant in love but not impassible, constant in faithfulness but not immutable.

This relation established between God and man in Jesus Christ constitutes Him as *the place* in all space and time where God meets with man in the actualities of his human existence, and man meets with God and knows Him in His own divine Being. That is the place where the vertical and horizontal dimensionalities intersect, the place where human being is opened out to a transcendent ground in God and where the infinite Being of God penetrates into our existence and creates room for Himself within the horizontal dimensions of finite being in space and time. It is penetration of the horizontal by the vertical that gives man his true place, for it relates his place in space and time to its ultimate ontological ground so that it is not submerged in the endless relativities of what is merely horizontal. Without this vertical relation to God man has no authentic place on the earth, no meaning and no purpose, but with this vertical relation to God his place is given meaning and purpose. For that reason it is defined and established as place on earth without being shut in on itself solely within its horizontal dimensionality. Unless the eternal breaks into the temporal and the boundless being of God breaks into the spatial existence of man and takes up dwelling within it, the vertical dimension vanishes out of man's life and becomes quite strange to him—and man loses his place under the sun.

All this talk about the vertical and the horizontal could be taken in a purely symbolic way or be construed as applying only to the relation of human existents to Being itself after the fashion of Martin Heidegger,[1] but then we would be right back in some form of dualism, deism or even atheism. What we need, however, is a satisfactory way of expressing in modern terms the fact that in the Incarnation the eternal reality of God has actually intersected with our creaturely reality, overlapping with it in Jesus Christ in a definite span of space–time, and thus constituting Him the one place where man on earth and in history may really know the Father because that is the place where God Himself has elected to dwell among us. The difficulty that faces us is that this span of space–time is a coordinate system of divine and human, eternal and temporal, invisible and visible, spiritual and material relations, and we want to coordinate them in one and the same language. *But that is exactly what we cannot do.* Yet it is because people keep on trying to do this that they continually introduce confusion into theology, and then because this inevitably breaks down they conclude that the Incarnation was not after all an actual intersection of divine reality and this-worldly reality, and so they lapse back once more into a false dualism substituting a *hoc significat* in place of *hoc est.*

The problem that confronts us here is not unknown in other areas of knowledge. Indeed it may be said that this is one of the chief problems facing natural science at the moment: how to develop unitary field theory in such a way that full weight is given to language expressing knowledge arising out of observational experience, such as quantum theory, and to

[1] Cf. Martin Heidegger, *The Question of Being*, tr. with introduction by W. Kluback and J. T. Wilde, 1959.

language arising out of theoretico-intuitive penetration into the connections that are not observable but no less real, such as relativity theory.[1] This is not to say that quantum theory can be expressed merely in language deriving from observational experience or that relativity can be expressed without relation to observational language, for they actually overlap, but there is a significant if relative difference in their languages in this respect. It should be noted, however, that when we develop the second type of knowledge and language out of field theories, the particulate view of nature that physics developed before then, by way of abstractions from observational knowledge, undergoes a decided change, and it is in that changed form that it must play a part in unitary field theory. It would appear that the linguistic philosophy in which so many are now immersed is a hang-over from that old particulate view of nature and is unable to cope with the kind of scientific connections that arise in field theory, and for that matter in theology.[2]

How may we express the space–time of the Incarnation as a coordinate system of real relations in such a way as to do justice both to the divine and to the human centres of reference, and therefore to coordinate the corresponding movements of thought and speech about them without confusing them? In what follows I wish to set forth the main lines of a unitary approach in which this may be done, drawing on the help of ancient and modern thought.

(1) We take the line adopted by the Church Fathers in their

[1] This has been illuminatingly discussed by Jean Charon, *Man in Search of Himself*, ch. 3, although he rather exaggerates the contrast between the two languages. I have taken up the problem he has posed in the intersection of the two languages, but cannot follow him in the solution he offers in 'a deeper conception of religion', ch. 6.

[2] This relation of critical and linguistic philosophy to the particulate view of nature has been powerfully demonstrated by Errol E. Harris, *The Foundations of Metaphysics in Science*.

understanding of Jesus Christ in space and time as God's place in this world where He is present in our place. This is not to be treated as merely a metaphorical way of speaking, for this place is not a vacuum but location in the context of real being, divine and human. Jesus Christ is the place of contact and communication between God and man in a real movement within physical existence, involving interaction between God and nature, divine and human agency. Without such a place within our created and historical existence where God has made room for Himself there could be no actual interconnection by word or deed between God and man. It is not in some ideal world but in our estranged world that God does this. It is place in which God has condescended to enter within the spatial context of the physical world in order to bring His boundless being to bear directly on man, so that we must think of that real presence not apart from the objective determinations and conditions of our physical existence. It is place in which God has condescended to live His divine life within the time of man's living experience, so that we must think of that living presence not apart from the temporal determinations and conditions of our historical existence. It is place that is filled with the energy of divine being and life, but place that is also filled with the energy of human being and life. Hence we have to think it out in a kinetic way in accordance with the nature and activity of God who locates Himself in our space and time as one with us, and in accordance with the nature and activity of the earthly and temporal existence in which He has become incarnate. Yet that earthly and temporal location cannot be so defined and delimited as to define and delimit the priority of God over created space and time, so that the place of God in Jesus Christ must be an open concept, rooted in the space and time of this world yet open to the transcendent presence of God. Put the other way round, this is to say that the transcendent God is present and immanent within this world in such a way that we encounter His transcendence in this-worldly form in Jesus Christ, and yet in such a way that we are aware of a majesty of transcendence in Him that reaches out infinitely beyond the whole created order.

Patristic theology worked this out in a relational and differential mode of thought in which it made use of the concepts of 'economic condescension' and 'hypostatic union'. By economy it referred to the orderly purpose and control of God introduced in and through the Incarnation in such a way that it was not imposed upon the world from without but operated from within it. Economy represents the great condescension of God the Son to work out the eternal purpose and love of God without violence within the alienated life of man and the disordered existence of the world. The mode of economic condescension is the way of Jesus Christ the Servant. In humble subjection and sacrificial passion He brought the holy power of God to bear upon human weakness and corruption by sharing in them Himself, thereby restoring to creaturely existence being and rationality against the downward trend of nature into non-being and irrationality. This represented a fulfilling and completing of the true movement of nature against itself, importing its redemption and re-creation, not its abrogation.

The intimate relation between God and the creation this entailed was concealed precisely by the entry of God within created existence and by the forces of enmity and decay entrenched within it, but it does become revealed in the person of Christ as Mediator between God and man, which the Fathers brought to expression in the concept of the 'hypostatic union'. This is the ontological counterpart to 'economic condescension', but even here the very fact that God Himself in His own being is present and at work, in abruptly divine acts comparable only to the Creation itself, limits our observing and apprehending of the inner relation between God and man in Christ. We cannot speak of it in such a way as to transgress the limits between the Creator and the creature, as if we could discern those processes by which in the Creation the observable comes into observation, and by which in the Incarnation the eternal Son enters into the perceptibility of the world. However, now that the Son of God has become man, appropriating to Himself perceptibility and conceptuality, together with linguistic communicability, from created existence, He confronts us men where we are in space and time, organizing the space–time structure of His human nature and life in relation to our existence in such a way that we are able to know the Father He reveals through Himself the Son, and are able also to say of the Man Jesus Christ that what He is towards us in divine acts of forgiveness, regeneration and sanctification, He is in Himself, antecedently, inherently and eternally. He is God of God, as well as Man of man.

When we reach this point in the acknowledgement of the real Being of Jesus Christ as divine as well as human we are forced to go back along the *a posteriori* road we took from His Humanity in order to know Him truly and adequately, still *a posteriori*, out of the integrity of His divine and human natures and in accordance with their unity in His Person. This amounts to a repetition in theological development of the actual way in which Jesus Christ was disclosed and known in the first place to the New Testament witnesses as Son of the Father, Himself God manifest in the flesh, without any detraction from the perfection of His creaturely and human nature among men. It was precisely this understanding that Nicene and Chalcedonian Christology sought to express in such a way that it did full justice to the intersection and overlapping of divine and human reality in Jesus Christ, yet in such a way as to reject any confusion or separation between them. Judged by modern scientific standards alone it was thus an exemplary model of unitary theory and of the way in which the languages of the observable and the non-observable are to be co-ordinated. Yet the Nicene-Chalcedonian Christology cannot be regarded as merely a theory so much as the organized form of apprehension and conceptualization forced upon the Church by the ontic necessity of the given Reality of God in Jesus Christ, although admittedly it cannot be confined within the concepts and statements used. Doubtless it stands in need of restatement and modification in view of the fuller knowledge of Christ that it has helped to mediate, but the very fact that all through the history of Christian thought its economic simplicity has proved so astonishingly fertile in deepening and enlightening our thought far beyond the range of its original application, can only command our profound respect.

(2) Another way to express the co-ordination of divine and human centres of reference in the space–time of the Incarnation seems to be offered by *the analogy of topological language* in which physicists seek to represent the difficult elastic connections between the dynamical and geometrical aspects of things or between quite different kinds of space. An attempt was made in a similar direction by some of the Greek Fathers to connect up the different ways in which we must speak about *topos* or *place* in accordance with the human and divine natures of Christ, physical *topos* and divine *topos*, and in which we must take into account a variational shift in the meaning and range of the concepts employed. The analogy used by John of Damascus for this purpose was that of 'mental place', τόπος νοητός, which he defined as place where mind is active and energizes and is contained not in a bodily but in a mental fashion. Apart from its special significance for him, in helping him to escape the toils of Aristotle's definition of place which

he unfortunately adopted, this served a double purpose: in contrast to Euclidean space, which is void of energy, it offered him a way of linking up physical space with divine space through the concept of energy or non-observable activity, and it enabled him to express the fact that the physical space of Christ on earth is open to passage beyond the limitations of the body. While He became incarnate within the physical space of the body He assumed, Christ was not confined or circumscribed by it. He thus became man without leaving the bosom of the Father, and while He became flesh He did not abandon His own immateriality.

This is rather like Whitehead's concept of the 'passage beyond nature' which he finds to be characteristic of mind. 'Mind is not in time or in space in the same sense in which events of nature are in time, but . . . it is derivatively in time and in space by reason of the peculiar alliance of its passage with the passage of nature. Thus mind is in time and in space in a sense peculiar to itself. . . . We all feel that in some sense our minds are in this room and at this time. But it is not quite in the same sense as that in which the events of nature which are the existences of our brains have their spatial and temporal positions'.[1] In a related line of thought Dr. Hugh Montgomery has spoken of the synthesis of the physical and mathematical approaches in which we are concerned with a kind of space that is both in and out of time, and related it to the transcending of static viewpoints in the contemplation of a great work of sculpture such as the Delphic charioteer. 'The impression one gains is one of eternity: his eyes are gazing out of time.'[2] This is precisely the point that Byzantine art often sought to incorporate in its representations of Christ when it reversed the

[1] A. N. Whitehead, *The Concept of Nature*, p. 69 f.

[2] H. Montgomery, from a paper entitled 'Space', to be published in *The Philosophical Journal*, vol. 6, No. 2, July, 1969.

natural perspective of the dais on which He was made to stand in order to open out the range of vision from the spatial and temporal into the infinite and eternal, from the material and visible into the immaterial and invisible Majesty of God Pantocrator who is transcendent over all.

Now although analogical elements are involved it is not with analogical but with a form of topological language that we are concerned here, in which realities or states that are not proportionate to one another, such as the observable and the non-observable, are yet meaningfully coordinated. This is done not through considering them merely in their particularities or disparateness, for then they could only be correlated in logico-mechanical ways, nor by considering them merely in the inter-connections set up by movement in some field which would be inconceivable apart from that field, but in a new kind of order which does not abrogate, although it may modify, the patterns of order that arise when they are considered apart. It is in this direction that we are pointed by relativity theory when it shows that what is observable cannot be represented with scientific precision without reference to what lies outside observation altogether, by penetrating into connections that are real although they can have no direct correspondence to our observational experience, and so by showing that a different kind of structure, that of space–time, is required if we are to think out the interrelatedness of things in the cosmos. It is characteristic of space–time structure that it is differently disclosed according to the field of force in view, and that it has sufficient differentiality to be able to absorb new patterns of order without the abrogation of others, and thus it opens out beyond the particulate patterns that arise in space and in time. Topological language can be used to express the relation of place in the physical sense, as spatial and temporal location, to the whole of space–time through the consideration of some field of energy or action, and through the extension of the kind of connection that emerges in it. It attempts to rise above the level of thought in which we think, as it were, in simple geometry of patterns of corpuscular distribution, to the level in which we think of distinguishable situations and positions of things within the structural diversities and continuities of fields of force. Unless we can make this transition to the level of thought in which these new kinds of structure are revealed we remain stuck at levels which deprive us of the ability to penetrate into those real connections which we require for precision in our knowledge, 'precision' meaning what is strictly in accordance with the actual nature of things and their inherent rationality.

This is similar to the problem with which biologists are faced, if they remain stuck in mechanistic concepts and try to account for living functions only in terms of the laws of physics and chemistry. What is needed is transition to the level of organismic thought in which a different kind of structure is discerned which does not abrogate the laws of nature expressed by physics and chemistry, but which is coordinated with them at what Polanyi calls their 'boundary conditions' which nature, in so far as we bring it to representation on the levels of physics and chemistry, has left open and indeterminate. The approach to this is not by way of reduction, for the damage by the specification of particulars may be irremediable, but through attention to the field set up by living force and the operational principles that are revealed in it.[1] Thus what obstructs and prevents topological modifications of our thought in these areas is the particulate view of nature, together with the very restricted concepts and the logico-mechanical structures that are thrown up as a result of its narrow and rigid dualism. Modern biology has yet to achieve

[1] Michael Polanyi, *The Tacit Dimension*, especially ch. 2.

a 'break-through' in this direction comparable with that in physics, but when it does, the organized space–time structures of the biological field should supply historical science and theological science with more apt analogues than those which are now available in physics.

It is once again basically the same problem that faces us in Christian theology, for example, in regard to the resurrection of Jesus Christ from the grave. If we approach that in the phenomenological manner, abstracting it from its place in the field of force set up by God in the Incarnation, thus isolating it from the whole interrelatedness of creation and redemption, to consider it only in its individual aspect as one phenomenon to be correlated with other phenomena according to the laws of physics and chemistry, then it is no wonder that we find it a stumbling-block. But if we refuse to abstract it from the field of living power disclosed in the Person of Christ, who is after all the Subject of the resurrection, and think of it not only in terms of the successive and coherent structures of His life and work on earth but in terms of His whole space–time track in the cosmos, we will be able to think of it in quite a different series of connections which would have been inconceivable outside that field, but which now thrust themselves upon us demanding recognition as the inner reality and rationality of the Incarnation. Moreover, the force that discloses to us this field of organized connection is the Holy Spirit and it is in accordance with His nature as the immediate personal energy of the Being of God, that we are able to discern the operational principles which, to use Polanyi's language, emerge at this level and control the boundary conditions left open by the operations at the lower level without abrogating them, namely, the sheer creativity of the living God. It is here that we see again at work the relation between the created rationalities and the transcendent rationality of God in which the latter is recognized not as an intrusion into the former but rather as their affirming and establishing on their true and ultimate ground.

I do not wish to suggest that we can master our difficulties of thought and expression by means of topology, as though we could extrapolate that from physics into theology, but rather that we must engage in a kind of topological modification of our concepts and forms not altogether unlike the modification which we find in the organized geometry of the physicists.

(3) Another line of approach to a way of speaking of the Incarnation as a coordinate system of vertical and horizontal dimensions seems to be opened up for us by Gödel's famous theorem about the undecidability of certain propositions within the rules of a formal system. We recall that our difficulty is that we are unable in one and the same language to speak consistently and unambiguously of the intersection of divine and human reality. The Church Fathers operated with a conscious ambivalence in their use of terms in respect of a differential relation in our thought towards God in accordance with His nature and towards man in accordance with his nature. The Medievals operated with the instrument of analogy in which they sought to clarify the status of ambivalence, acknowledging the lack of proportion between God and man while preserving the connection between them without which human thought and speech of God would be meaningless. However, since our concepts and statements are limited by this world and necessarily have creaturely content, they may be used analogically in such a way only that we intend God beyond their creaturely content. This contained a serious difficulty that troubled Duns Scotus. He held that the intellect

[1] See *De doctrina Ioannis Duns Scoti* (ed. by P. Balić, Rome, 1968), vol. IV, pp. 293 ff., 298 ff.

assents to a complex of propositions as true, and not only as logically valid, provided that at least one basic proposition is immediately self-evident to the intuitive apprehension. But we cannot have an intuitive knowledge of God as an actually present object, immediately and evidently grounded upon it, for that would take us beyond the condition of this present life—yet if our analogical statements about God are to be true there must be at least one in which the reality of God bursts through the analogical form by which we intend Him and is disclosed in His own self-evidence in distinction from it or apart from it. Duns Scotus' answer to this was to distinguish between different gradations of knowledge which are co-ordinated through one another to the supreme level in which God knows Himself and is His own completely adequate and proportionate Object. Through that coordination between the levels God remains for knowledge at all levels its one proper object. In that translogical or metaphysical relation concepts referring to God are analogical, but when we consider terms on one and the same level in their second intentions and logical interrelations they must be regarded as univocal. The complex of propositions on any level must be incomplete in order to be true, so that the hierarchy of levels would hold good only if they were made open at the top by God by being anchored in His own self-knowledge.

Although some people in the past have held that there are different metaphysical levels in our knowledge of God and the world, no one seems to have followed Duns Scotus in his treatment of the analogical reference of our concepts and terms to God through a coordination of the different levels in which those concepts and terms are employed logically. But this way is now open for us since Gödel has demonstrated, with reference to the *Principia Mathematica* of Russell and Whitehead, that no such logico-deductive system can be complete and consistent at the same time, and that it can be consistent only if it is incomplete—any system that was both complete and entirely self-consistent would be a meaningless game.[1] That is to say, Gödel showed that in any formal system there must be some propositions which cannot be decided, proved or disproved, within that system, if it is consistent. If they are to be proved or disproved that must be done through integrating them in another and wider formal system. But this means that we must operate with different levels of formal systems, whether logico-deductive or syntactical, each of which is connected with the other through having certain propositions in common, and that any formal system has meaning only through semantic reference beyond itself. It must be added that this openness or incompletability of a formal system does not arise out of any inability in the human mind, but out of the very nature of things, so that distinctions in metaphysical levels are also involved. Gödel's theorem applies, of course, not only to logico-deductive systems such as arithmetic but to syntactical systems that are not thrown into logico-deductive form. Even logico-deductive systems must be coordinated with the level of ordinary language, if they are to have any applicability to existence—otherwise they are no more than games and can have no serious bearing upon science.

When this is applied to our problem in the ambiguity of our language, it brings immense clarification. Consider the difficulty of the physicist in developing his understanding of space and time by advancing his thought to higher levels with new concepts and connections which do not correspond with our observable experience. On the advanced level of his

[1] Kurt Gödel, 'Über formal unentscheidbare Sätze der *Principia Mathematica* und verwandter Systeme I', *Monatshefte für Mathematik und Physik*, 38. Band, pp. 173–98, Leipzig, 1931; E.T. by B. Meltzer, Oliver and Boyd, 1962. Cf. also R. Carnap, *Logische Syntax der Sprache*, pp. 93 ff. Vienna, 1934.

thought the terms space and time are defined within the formal calculus he has developed, and therefore there can be no one-to-one correspondence with space and time as defined within the system of ordinary language, but they must be co-ordinated through the hierarchical structure that connects the different levels, otherwise they would be completely detached from our ordinary experience and could have no meaning for us at all. Nevertheless, the new concepts of space and time as developed in the symbolic language of the physicist could not be described in ordinary language, without confusion and contradiction, since the terms for space and time in each language-system are differently defined within it, even though they may be linguistically identical. Thus it holds that each formal system is open upwards, not downwards, and therefore no level of thought can be accounted for by being reducted downwards—which of course links up closely with our discussion of topological language.

All this applies with equal force to theological science in its relation to natural knowledge and to ordinary or natural language. Here then we have a way of coordinating the different languages about God and man in such a way that they are not confused or identified with each other. It is this coordination that gives ground and objectivity to our proper analogical references, without our having to read their creaturely content into the divine Being. But while the different levels of our thought are open upward in this way to God, they are nevertheless to be coordinated downward with basic statements arising out of our ordinary experience and empirical knowledge. That is why theological formulations cannot be without their empirical correlates. This does not mean that each theological statement must have such a correlate, but that it must be embodied in a system in which some basic statements do have correlates on the empirical level, otherwise they would be entirely without meaning and applicability in our human and creaturely existence. Hence it is quite impossible for Christian theology to be indifferent to the questions of historical facticity, for any construct of Christ that has no rooting in actual history can only be a vehicle of our fantasies. Thus, for example, the Christian doctrine of the resurrection cannot do without its empirical correlate in the empty tomb; cut that away and it becomes nonsensical.

It is not only with ordinary experience, however, that theological statements must be coordinated, but with each science that is concerned with the examination and interpretation of this-worldly experience and knowledge. Hence theological science must examine the empirical correlates of its own basic statements in discussion with the other sciences whose operations overlap at these points with its own, testing those empirical correlates for their truth or falsity, yet in such a way as not to allow the real *decision* (in Gödel's sense) to be taken outside the organized structures and 'topological' connections within which it must operate as a distinctive science. While some form of 'dual control' (in Polanyi's sense) is envisaged here, the final decision must be taken within the organizing principles of the higher level over those statements which it shares with the level below it and which are not absolutely decidable within it. But since no level of thought and no science can be a completely closed system the decisions must be taken in the light of the way in which the various sciences can be coordinated meaningfully in levels in which they can best serve each other in the whole of our knowledge of God and of the world He has made.

INDEX